VGM Opportunities Series

OPPORTUNITIES IN MARINE AND MARITIME CAREERS

Wm. Ray Heitzmann, Ph.D.

Foreword by
Jean-Michel Cousteau
Director
Jean-Michel Institute

VGM Career Horizons
NTC/Contemporary Publishing Group

Library of Congress Cataloging-in-Publication Data

Heitzman, William Ray.
    Opportunities in marine and maritime careers / Wm. Ray Heitzmann :
Foreword by Jean-Michel Cousteau.—Rev. ed.
       p. cm.—(VGM opportunities series)
     ISBN 0-8442-1833-2.—ISBN 0-8442-1838-3 (pbk.)
     1. Merchant marine—Vocational guidance—United States. 2. Naval
art and science—Vocational guidance—United States.
3. Oceanography—Vocational guidance—United States. 4. Fisheries—
Vocational guidance—United States. I. Title. II. Series.
VK160.H37   1999
387.5'023'73—dc21   98-40925
                                                       CIP

Cover Photo Credits:
Images copyright © 1997 PhotoDisc, Inc.

Published by VGM Career Horizons
A division of NTC/Contemporary Publishing Group, Inc.
4255 West Touhy Avenue, Lincolnwood (Chicago), Illinois 60646-1975 U.S.A.
Copyright © 1999 by NTC/Contemporary Publishing Group, Inc.
Printed in the United States of America
International Standard Book Number: 0-8442-1833-2 (cloth)
                                         0-8442-1838-3 (paper)

99 00 01 02 03 04 MV 18 17 16 15 14 13 12 11 10 9 8 7 6 5 4 3 2 1

DEDICATION

To My Family: Past, Present, and Future.
And to the Marine/Maritime Family Everywhere.

CONTENTS

ABOUT THE AUTHOR

Wm. Ray Heitzmann, Ph.D., a faculty member at Villanova University, has a national reputation in the area of marine and maritime studies. His articles on these topics have appeared in *Sea World, Shipmate, Sea History, Sea Heritage News, The U.S. Naval Institute Proceedings, The Beachcomber, The Journal of Marine Education, Naval History, Long Beach Island Magazine, American Neptune,* and others. He also has run career education workshops and courses and has published related articles in *Career World, Real World,* and the *California Social Studies Review.* For several years he authored a column in *Long Beach Island Magazine* entitled "Nautically Notable and the World of Water."

Ray Heitzmann has served as a high school teacher and as a basketball coach in Illinois, New York, and Pennsylvania; he is the former head men's basketball coach at Neumann College in Pennsylvania. Following undergraduate work at Villanova University, he received his master's degree from the University of Chicago and his Ph.D. from the University of Delaware. He has taken additional graduate studies at California State University at San Jose and Northwestern University in Illinois.

FOREWORD

Together with my father, Jacques Cousteau, I have been a witness, often firsthand, to humankind's entry into the sea as never before—on SCUBA as menfish, in small maneuverable submarines, and as aquanauts living day after day in the liquid world without sun. As we have entered the ocean in greater numbers and more diverse ways, we have learned much about the sea and about ourselves.

As a young architect, my dream was to design underwater cities. I now realize this will only come to pass as an exotic oddity—an occasional underwater hotel or science research station—because, even with the wonders of our technology, we are still aliens underwater, who are able to visit in ecstatic excursions but unable to settle.

Though the sea has taught us of our limitations, it has also expanded our vision, as we watch through film the many forms life has taken and the ways in which fish and worm and whale have evolved to survive. The study of a coral reef is a description of a large city in which competing and cooperating races and professionals and individuals live and thrive and die in a balance that allows the reef to continue for thousands and thousands of relatively undisturbed years. Much of what we have learned at sea is a parable of what we could do on land.

But watching, smelling, feeling, tasting, and thinking about the sea did more than teach me—it changed me forever into a man in love with the sea, just as my father has been and his father before him. We quickly learned that people protect what they love, and we are in love with the sea.

In reading this informative book and considering whether you want an ocean career, you will find you have many choices; some of them will exhilarate you, and some will disappoint you. Choose wisely and base your choice on what you love. Few go to sea to become rich. Almost no one comes back from the sea who isn't enriched as a human being. Be practical about your choices and spend all the time you can on every rocking boat, mudflat, or laboratory. See for yourself how the sea suits you. Then love her with passion and save her through your work.

Jean-Michel Cousteau
Director
Jean-Michel Institute
Santa Barbara, CA

PREFACE

THE WORLD OF WATER

When the first edition of this book was written more than twenty years ago, there existed a great need to remind Americans of their rich maritime traditions and the vitally important role water played in their lives. That need still exists. In fact, our neighbors to the north seem more in touch with the marine world than we do.

Although boarded by eleven thousand miles of coastline, many Americans behave as if the United States is a landlocked nation. The marine and maritime enterprise has become so ingrained in our traditions that we rarely give it much thought. For example, much of our language comes from nautical terms: words like "first rate," "shipshape," "blubber," "ironclad," "the whole nine yards" and "windfall" demonstrate how sea terms have entered into land use. Our economy, architecture, religion, music, and literature all bear the stamp of the sea.

Our planet, of which approximately 75 percent is covered by water, has been served by the sea. Early societies lived near rivers, on islands, or along oceans from which they obtained food, transportation, and trade—in a word, survival.

ACKNOWLEDGMENTS

The comprehensive nature of a book on marine and maritime careers requires the cooperation of a number of individuals and organizations. The following deserve special mention: Ms. Regina Agrusa, Carle Place, Long Island, New York; Ms. Andrea Green Bry, Ocean View, New Jersey; Ms. Teresa Wojcik, Council Rock School District, Pennsylvania; Ms. Kelly Green Yoh, The Haverford School, Pennsylvania; Ms. Nora Deans, Director, Publications, Monterey Bay Aquarium, California; Ms. Margaret Buchholz, Publisher-Editor, *The Beachcomber,* Ship Bottom, New Jersey; Ms. Joy Wolf, Director, Education Department, Sea World of California; Ms. Susan Alpert and Ms. Carolyn Stone, Public Relations, Cunard Line; Mr. David D. Beach, Manager, Technical Services, National Marine Manufacturers Association; Mr. Thomas V. MacEwen, Vice President, Prudential Lines; Mr. Walter Oates, Public Affairs Officer, Merchant Marine Administration, Washington D.C.; Dr. Chris Harrison, Dean, Rosenthel School of Marine and Atmospheric Science, University of Miami; Ms. Gretchen Waterbury, Chapman School of Seamanship, Stuart, Florida; Mr. Rick Tinnin, Marine Science Institute, University of Texas; Mr. Jeffrey A. Smith, Vice President, Public Affairs, The American Waterways Operators, Inc.; Mr. Les Bartholow, Office of Maritime Labor and Training, United States Department of Transportation; Mr. Enno Reckendorf, Director, The Norfolk School of Building; Ms. Nancy Richardson, Girl Scouts of the United States of America; Hon. Curt

Weldon, U.S. House of Representatives (Pennsylvania); Dr. James Merrill, Maritime Historian (Emeritus), University of Delaware; Ms. Mary Heitzmann Kingwill, Media/Technology Specialist, Weehawken Schools, New Jersey; Ms. Sharon Meeker, Sea Grant, University of New Hampshire; Dr. Nicole Crane, Monterey Peninsula College (California); Dr. Lundie Spence, Sea Grant, North Carolina State University/University of North Carolina; Mr. F. A. Courtenay, Jr., National President, the Propeller Club of the United States; G. Linda Crisculo, Washington Township High School, New Jersey; Mr. K. T. Weaver, Executive Director, United States Naval Sea Cadet Corps; Dr. Val Chase, Biologist, The National Aquarium in Baltimore; Professor Ken Stibolt, Anne Arundel Community College; Glenn S. Gordiner, Research Associate, Mystic Seaport Museum; Dr. Bill Hall, Marine Education Specialist, the College of Marine Studies, University of Delaware; Mr. Bill Ward, Education Director, Independence Seaport Museum; Mr. Rod Koon, Director of Port Relations, Port of Tacoma, Washington; Mr. John Wiegman, Assistant Director, Seafarers Harry Lundeberg School of Seamanship; Major Ken Conklin, Vice President, Seafarers Harry Lundeberg School of Seamanship; Mr. Robert E. Banks, Director of Marketing, South Jersey Port Corporation; Mr. Arthur W. Kane, Vice President, Corporate Relations, Norwegian Caribbean Lines; Ms. Liz Love, Monterey Bay National Marine Sanctuary, California; Mr. James A. Barber, Jr., Executive Director, United States Navy Institute; Mr. Peter Stanford, Editor-Publisher, *Sea History;* Ms. Bonnie Bekken, Social Studies School Service; Lt. Matt Zurcher, NROTC, Villanova University; Mr. Thomas Greene, Director Marine Science, Fort Hamilton High School, New York City; Professors Landry, Davis, Seitz, and Wilkie of Texas A & M University at Galveston; Robert Fraunberger, Bryn Mawr, Pennsylvania; Wil Tolen, Tolen Marine, Paducah, Kentucky; Ms. Lee Stack, Villanova University; Mary Heitzmann, University of Delaware; and Rick Heitzmann, Harvard University.

INTRODUCTION

The world of water offers an incredible number of exciting and rewarding careers that make an important contribution to our society. People in these jobs draw satisfaction not only from the personal enjoyment of their tasks, but also from the fact that they help others. Consider the following possibilities:

- You're a scientist and your current specialty has you investigating shipworms! These critters regularly devastate wooden vessels, piers, and pilings, costing millions. Your research has made an important contribution to controlling this "termite of the sea."

- You have studied underwater archaeology, and now you and your team have located and explored a shipwreck off the coast of Texas in the Gulf of Mexico. The expedition has brought to the surface artifacts that will provide insights to the historical era during which the ship operated. You have spent many months examining historical documents and old maps, securing government permission, and arranging for the latest technology to aid your search, and it has paid off.

- Here they come! You're entering your tenth year as a marine education specialist, and today you will be taking a group of sophomores from a local high school out on the Long Island Sound. The students first will be given a lecture on safety afloat and then some information on the bay. The students plan to take water samples and conduct some experiments back at school. You know it will be a long day, but you realize that the kids will love it and leave with a better understanding and appreciation of the marine environment.

- Your ship floats across the ocean as you stand on the deck, the wind blowing in your face. The freighter carries a cargo of heavy agricultural machinery to the west coast of Africa, where upon docking you will spend a few days before returning with a cargo of materials for American industry.

- Mayday! Mayday! Your Barnegat Light Coast Guard station has just received an SOS distress call and you have been ordered to rescue a sinking twenty-two-foot pleasure boat. Those aboard the small craft were enjoying a nice Sunday afternoon cruise when a sudden, serious storm hit their boat, imperiling their lives. Aided by your speeding boat and equipment, you arrive in time and successfully rescue the boaters. As you return to shore and receive the praise of your supervisor and the thanks of those saved from a watery death, you feel proud of your chosen profession.

- Aquaculture (fish farming) has had great success and promises an even greater future. It's clearly one of the most exciting projects in New Jersey's cities in years, and you're the marine science advisor. The plan is to raise fish (Nile perch) in deserted factories and warehouses. Once the fish reach marketable size, they will be shipped to fish markets on the East Coast. This aspect of aquaculture has great potential.

- The beaches in your state have suffered badly from what the media call the "El Niño" effect. The governor has asked you to provide technical skills in a project that will replenish the beaches and prevent future problems to the coastline.

- Your life as a marine biologist has led you to California. You are studying the growth of giant kelp at a test farm located about five miles off the coast of Corona Del Mar. The test farm is anchored in five hundred feet of water. The brown seaweed grows only in water depths of one hundred feet or fewer because of the need for sunlight. Upon harvesting, the experimental kelp will produce gas through a special process, thus alleviating, in part, the world's reliance on oil. You enjoy the excitement of your work and your potential contribution to the well being of the nation and the world.

- One of the fish species living in the streams of your state has contracted a disease. Because of the fear that the disease will spread, you and your supervisor, a fish culturalist, have traveled around the state, examining fish. Specifically, in your capacity as assistant, you have taken blood samples and analyzed them to see if the disease has spread to fish in other ponds or streams of the state. Each day you arrive back at your motel room tired but happy with your chosen job.

- While pursuing community college courses in marine propulsion technology, you work part time in a local marina. Because of the scarcity of trained people, you can work as much as you wish. However, your laboratory courses (particularly Marine Ignition and Fuel Systems) consume much of your time. At the end of two years, you will receive your Associate in Science degree. From several job offers, you've narrowed your choice to two —assistant marina manager or marine diesel technician for a local fishing fleet. Not only does your future look financially sound, you will have the satisfaction of living and working near the water.

- The great interest in the world of water has created a hunger for information on maritime and marine topics. You have always enjoyed writing. Now you write a syndicated monthly column entitled "Nautically Notable: The World of Water" for the general public.

- Fish farming in Idaho? Yes! In Buhl, Idaho, a company is developing a fish without bones. Believe it or not, millions of pounds of fish a year are raised through aquaculture! The Idaho project of which you're the director shows outstanding promise.

Do the above scenarios seem intriguing, challenging, and in tune with your career interests? If so, please read on to find the strategies you can pursue to achieve job satisfaction in a marine or maritime career. Just as with other occupations, people in marine and maritime careers spend many hours a day in routine detail work that is necessary for the success of any task. For this reason, many employers want people with specific marine or maritime abilities as well as good habits, such as punctuality, and writing and speaking skills.

This book provides information on several marine and maritime careers. The reader is strongly encouraged to write to and/or check web sites for additional information on jobs that appear interesting. Many sources of information are cited in the text and appendices. The schools mentioned herein are used as examples; discussion of them does not constitute an endorsement.

CHAPTER 1

AQUARIA-RELATED CAREERS

My job is varied. I enjoy the opportunity to interact with visitors of all ages and interests. Throughout a typical year I will:

- teach student, member, and general public programs
- develop and write lessons
- coordinate workshops for teachers
- organize special events
- train and supervise summer employees and interns
- work on our web page
- plan and lead trips to various locations (Florida keys, Massachusetts coast)
- attend conferences

—Gretchen Crapa

Gretchen holds one of the most desirable jobs in the world of water—she works at an aquarium, the National Aquarium in Baltimore. "I have always felt a deep affinity for the ocean and have been interested in marine creatures since I was a child. I have wonderful memories of spending hours at the touch pool at the Roanoke Aquarium." Undoubtedly, Gretchen was one of those visitors who asked, "How can I get a job at an aquarium?" This is one of the most frequently asked questions in the marine world today.

TYPICAL CAREERS

Several types of jobs exist at aquaria—some people work with animals, some with the public, some with both, and still others do none of these.

Gretchen began her career in the middle of a desert! Attending college at the University of Arizona (Sonoran Desert) she discovered "marine education" through a program that brought live animals, slides, and artifacts from the Sea of Cortez to schools in the Tucson, Arizona, area. Observing the excited faces of the children prompted her to continue to explore the marine world. After a summer internship at the Monterey Bay Aquarium, she became fully committed to her career. While currently employed in the education department of the aquarium, she continues to grow by pursuing a master's degree in environmental science and policy.

A variety of positions exist at most aquaria; many workers are employed behind the scenes and rarely work with animals or have contact with visitors. *Plumbers, artists, engineers, housekeepers, electricians, accountants,* and others contribute to the success of the facility. Some careers more commonly associated with the National Aquarium in Baltimore follow:

Aquarists maintain all aspects of the aquatic exhibits; this includes maintaining water quality, feeding the fish, cleaning the tanks, and completing related tasks including precise record keeping. People in these positions have backgrounds in marine or aquatic science, including SCUBA certification.

Aviculturists work with the marine birds and maintain all aspects of the display. This aquarium is unique in possessing such an exhibit. People employed in this specialty have a background in biology or zoology and have excellent public speaking skills.

Educators at the aquarium give presentations to visitors and teach classes. This requires an excellent knowledge of the various fish and birds and quality speaking skills. These individuals have backgrounds in science, usually marine science or biology.

Herpetologists manage the reptile and amphibian collection. They feed and care for these species. Again a background in the biological sciences is a must as well as close attention to detail. As with all animals at the aquarium, this collection must be carefully observed to maintain optimum health.

A number of other specialties remain critical to the success of an aquarium:

Veterinarians and their assistants, *veterinary technicians,* oversee the health of the animals. Veterinarians hold a medical degree (DVM) in animal medicine; veterinary technicians graduate from two or four-year colleges and pass licensing examinations. *Water quality technicians* ensure the quality and safety of the water environment.

Marine Mammal Trainer is perhaps the most popular and most sought after aquaria-related career. Armed with a background in psychology and/or biology, fine public speaking skills, and SCUBA certification, marine mammal trainers work with dolphins, seals, and orcas (killer whales). Though often visible as part of the "shows," these individuals labor many hours behind the scenes in training activities. They also help to maintain the tanks and the exhibit area. If this sounds like show business, perhaps it is.

Consider the career of Julie Scardina, animal ambassador and curator of animal training at Sea World of California. A weekly contributor to *Jack Hanna's Animal Adventures* and recent guest on *The Tonight Show* with Jay Leno, *Live with Regis and Kathy Lee,* and others, she has the skills and confidence to appear before millions on television. In preparation for training and performing with the eight thousand-pound Shamu, sea lions, otters, and walruses, she obtained her college degree from San Diego State University with a major in psychology and a minor in biology and then earned an associate's degree in exotic animal training and management from Moorpath College in California. Her career permits her both to interact with some of the world's most fascinating teachers and to provide education to children and adults. Clearly this occupation provides personal benefits as well as contributions to our society. For additional information contact:

International Marine Animal Trainers' Association
    John G. Shedd Aquarium
    1200 South Lake Shore Drive
    Chicago, IL 60605
    (312) 939-2426

Marine mammal scientists work closely with marine mammals; although some work at aquaria, most work in other environments. Employed by organizations ranging from marine laboratories to the U.S. Navy to the fishing industry, they study the behavior of a wide variety of marine animals. Despite the allure of this career, opportunities remain limited. Consequently, a strong background in science with very good grades are a must. Many have gained experience while serving as volunteers at aquaria and/or interns with private companies or government organizations. Most have graduate degrees and specialize, that is, they confine their study to a particular marine mammal—dolphin, manatee, and so forth.

The following web site contains additional information:

    http://www.une.edu/marbio/marmam.html

Visiting and volunteering at aquaria or oceanaria can serve as valuable career preparation. Technically, aquaria emphasize marine ecosystems involving all species including marine mammals. They fall into the following categories: publicly owned or governmentally owned, not for profit, and privately owned for profit. Oceanaria are privately owned for profit, such as Sea World. The latter tend to emphasize marine mammals.

Aquaria tend to focus on special exhibits, some of which change periodically. Canada's Vancouver Aquarium has had a "Whale Fest," the New York Aquarium (Brooklyn) boasts a "Bermuda Triangle" that emphasizes marine animals (sea turtles, moray eels) from that region, the Aquarium of the Americas (New Orleans) has a wonderful Mississippi Delta exhibit, the North Carolina Aquarium at Roanoke Island features a freshwater "Turtle Pond," and the Maritime Aquarium at Norwalk integrates the marine and maritime worlds.

AQUARIA AND OCEANARIA

The following offer wonderful opportunities:

Aquarium for Wildlife
 Conservation
Boardwalk and West Eighth
 Street
Brooklyn, NY 11224
(718) 265-3405

Aquarium of Niagara Falls
701 Whirlpool Street
Niagara Falls, NY 14301
(716) 285-3575

Aquarium of the Americas
1 Canal Street
New Orleans, LA 70130
(504) 565-3033

Aquarium of the National Marine
 Fisheries Service
166 Water Street
Woods Hole, MA 02543-1097
(508) 548-5123, ext. 267

Belle Isle Aquarium
P.O. Box 39
Royal Oak, MI 48068-0039
(313) 267-2727

Bermuda Aquarium, Museum,
 and Zoo
P.O. Box FL 145
Flatt's FL BX Bermuda

(441) 293-2727

Clearwater Marine Science
 Center Aquarium
249 Windward Passage
Clearwater, FL 34630
(813) 441-1790

Dallas Aquarium
P.O. Box 150113
Dallas, TX 75315-0113
(214) 670-8453

John G. Shedd Aquarium
1200 South Lake Shore Drive
Chicago, IL 60605
(312) 939-2426

The Living Seas
P.O. Box 10000
Lake Buena Vista, FL 32830
(407) 560-7688

Marine World-Africa USA
Marine World Parkway
Vallejo, CA 94589
(408) 648-4800

Marineland of Florida
9507 Ocean Shore Boulevard
Marineland, FL 32086
(904) 471-1111

The Maritime Center at Norwalk
10 North Water Street
Norwalk, CT 06854
(203) 852-0700

Miami Seaquarium
4400 Rickenbaker Causeway
Miami, FL 33149
(305) 361-5705

Monterey Bay Aquarium
886 Cannery Row
Monterey, CA 93940-1085
(408) 648-4800

Mount Desert Oceanarium
172 Clark Road
Southwest Harbor, ME 04679
(207) 244-7330

Mystic Marinelife Aquarium
55 Coogan Boulevard
Mystic, CT 06355-1997
(860) 572-5955

National Aquarium
U.S. Department of Commerce
Building
Fourteenth and Constitution
Avenue, NW
Washington, DC 20230
(202) 482-2825

National Aquarium in Baltimore
Pier 3, 501 East Pratt Street
Baltimore, MD 21202-3194
(410) 576-3800

New England Aquarium
Central Wharf
Boston, MA 02110
(617) 973-5220

New Jersey State Aquarium at
Camden
1 Riverside Drive
Camden, NJ 08103-1060
(609) 365-3300

North Carolina Aquarium at Fort
Fisher
2201 Fort Fisher Boulevard
South, Box 1
Kure Beach, NC 28449
(910) 458-9825

North Carolina Aquarium at Pine
Knoll Shores
P.O. Box 580
Atlantic Beach, NC 28512-0580
(919) 247-4004

North Carolina Aquarium on
Roanoke Island
P.O. Box 967, Airport Road
Manteo, NC 27954
(919) 473-3494

Ocean World Inc.
1701 Southeast Seventeenth
Street
Ft. Lauderdale, FL 33316
(305) 525-6612

Oregon Coast Aquarium
2820 Southeast Ferry Slip Road
Newport, OR 97365
(541) 867-3474

Sea Life Park Hawaii
41-202 Kalanianaole Highway,
Suite 7
Waimanalo, HI 96795
(808) 259-7933

Sea World of California
1720 South Shores Road
San Diego, CA 92109-7995
(619) 222-6363

Sea World of Florida
7007 Sea World Drive
Orlando, FL 32821-8097
(407) 351-3600

Sea World of Ohio
1100 Sea World Drive
Aurora, OH 44202
(216) 562-8101

Sea World of Texas
10500 Sea World Drive
San Antonio, TX 78251
(210) 523-3000

The Seattle Aquarium
1483 Alaskan Way,
Pier 59
Seattle, WA 98101-2059
(206) 386-4300

South Dakota Discovery Center
and Aquarium
805 West Sioux Avenue
Pierre, SD 57501
(605) 224-8295

Steinhart Aquarium
Golden Gate Park
San Francisco, CA 94118-4599
(415) 750-7247

Stephen Birch Aquarium-Museum
Scripps Institution of
Oceanography
University of California, San
Diego
La Jolla, CA 92093
(619) 534-4086

Tennessee Aquarium
P.O. Box 11048
Chattanooga, TN 37401-2048
(800) 262-0695

The Texas State Aquarium
P.O. Box 331307
Corpus Christi, TX 78463
(512) 881-1200

University of Puerto Rico
Department of Marine Sciences
Museum
P.O. Box 5000
Mayaguez, PR 00681-5000
(809) 832-4040, exts. 3443,
3447, and 3838

Vancouver Aquarium
P.O. Box 3232
Vancouver, British Columbia
Canada V6B 3X8
(604) 685-3364

Waikiki Aquarium
2777 Kalakaua Avenue
Honolulu, HI 96815
(808) 923-9741

The following organizations active in the marine world can provide additional information:

American Fisheries Society
5410 Grosvenor Lane
Suite 110
Bethesda, MD 20814

American Society of
Mammologists
Institute of Ecosystem Studies
Box AB
Millbrook, NY 12545

American Society of
  Mammalogists
Virginia Museum of Natural
  History
1001 Douglas Avenue
Martinsville, VA 24112

Center for Coastal Studies
Box 1036
Provincetown, MA 02657

Cetacean Behavior Laboratory
Department of Psychology
San Diego State University
San Diego, CA 92182

Coastal Ecosystems Research
  Foundation
2173 West Sixth Avenue, #207
Vancouver, British Columbia
Canada V6K 1V5

EarthWatch
680 Mount Auburn Street
P.O. Box 403
Watertown, MA 02272-9104

Kewalo Basin Marine Mammal
  Laboratory
1129 Ala Moana Boulevard
Honolulu, HI 96814

Marine Mammal Research Group
EPCOT Center Trailer #251
Walt Disney World Company
P.O. Box 10,000
Lake Buena Vista, FL 32830-
  1000

Marine Mammal Research
  Program
Texas A&M, University at
  Galveston

Department of Fisheries and
  Wildlife
4700 Avenue U, Bldg. 303
Galveston, TX 77551

Marine Science Association
Biology Department, Box
  18540
Jackson State University
Jackson, MS 39217

Mingan Island Cetacean Study
285 Green Street
Lambert, Quebec
Canada J4P 1T3

National Oceanic and Atmospheric
  Administration
1335 East-West Highway
Silver Spring, MD 20910

National Wildlife Federation
1400 Sixteenth Street, NW
Washington, DC 20036

Pacific Whale Foundation
Kealia Beach Plaza
101 North Kihei Road, Suite 21
Kihei, HI 96753-8833

Theater of the Sea
P.O. Box 407
Islamorada, FL 33036

U.S. Fish and Wildlife Service
1001 East Tudor Road
Anchorage, AK 99503

Whale Research Group
230 Mount Scio Road
Memorial University of
  Newfoundland
St. John's, Newfoundland
Canada A1C 5S7

Women's Aquatic Network
  Box 4993
  Washington, DC 20008
  http://orgs.womenconnect.com/wan/

Consult also the ":Whale Net" web site at http://whale.wheelock.edu/

CHAPTER 2

CRUISE SHIP CAREERS

It's glamorous. It's fun. It's demanding, hard work.

The cruise ship industry has experienced dramatic growth with seven million passengers currently cruising and revenues of close to $10 billion!

Jobs are plentiful, and if you enjoy traveling, meeting people, and going to exotic destinations, this may be your dream job. A wide variety of careers exist, and almost every major city in North America can boast of passenger ships ready to depart. A warning: Little downtime exists for employees and many demands must be met onboard a ship. Most individuals work for two to four years, have a great time, and then "retire" to employment ashore.

REQUIREMENTS

Those wishing to work on a passenger ship must have background and experience in a specific job. For example, servers, aerobics instructors, cosmetologists, and wine stewards all must have training and experience on land before ever boarding a ship. Don't despair; some jobs require only low-level skills that can be easily learned. In addition to training and experience, there are a number of other important requirements, starting with excellent health. Most ships will have a medical doctor onboard to assist guests and crew in case of emergency. However, the medical care is minimal, and employers don't want employees who have frequent illnesses resulting from the demands of work aboard ship. Although age is generally not a factor, people in their twenties and thirties have an edge; a demand does exist for those in their forties, fifties, and sixties for specific jobs.

The ability to speak a foreign language will increase the probability of obtaining a job; many non-English speaking tourists are on ships, and a foreign language will help you in your job search on foreign vessels.

Two personality traits are extremely important to success aboard ship—flexibility and dependability. Flexibility is important because a person's lifestyle will change drastically. Dependability is equally important. Although life aboard ship appears glamorous—it is for the guests—it is not necessarily glamorous for the staff! Too often those who labored to obtain a cruise ship job quit after a few months—approximately 25 percent of first-time employees do not fulfill their contractual commitment (usually one or two years). Consequently, reliability ranks as an important personal characteristic.

CAREERS

Aboard ship two categories of careers exist: those dealing with the ship and those related to the cruise-hotel aspect of the pleasure vessel.

The cruise ship has officers and a crew concerned with its navigation, maintenance, propulsion, and so on. This would be comparable to any maritime vessel (see Chapter 6). The routes to these jobs would be through the merchant marine academies and the training programs of the various unions. Also, some former naval officers and crew may land some positions in these areas.

The following are some of the jobs related to the luxury hotel–cruise type careers; all of these careers require the ability to relate well to people and to assist the guests in having a good time.

Bartenders prepare drinks at the various bars located throughout the ship. Experience is a must, as are details of the latest "in" drink as well as the score and details of the previous night's game.

Cosmetologists and *barbers* perform the hairdressing necessities for the passengers. Normally these individuals have excellent skills and often have reputations in their field.

Cruise directors supervise the staff to maximize the enjoyment of the passengers. For example, a director will arrange for tours when the ship stops at the various ports during the cruise. It is often the director who handles most of the problems that arise for passengers. Persons wishing a position as cruise director would list extensive travel experience and tenure as an *assistant director.* Often the success of the cruise is directly related to the competence of the director.

Ship's photographers offer passengers the opportunity to remember their voyage through photographs. This busy position (some larger ships have two) requires travel ashore and attendance at parties and special events. Skills involved in this position are those required of any quality professional photographer. If you would like to inquire about this career, contact the following:

TransOcean Photos
New York Cruise Ship Terminal
Suite A
West Fifty-fourth Street and Twelfth Avenue
New York, NY 10019

Sales personnel aboard ship work in places such as the gift shop. This is often a concession on a ship, so those seeking employment must apply to the appropriate concession company. Obviously retail experience is a must for this job.

Wine stewards are the wine experts aboard ship. They answer passengers' questions concerning the beverage—the best wine to drink with certain foods and the correct procedures related to wine drinking. Often these individuals have attended specific schools to obtain their expertise.

Chief pursers, as may be gathered from the name, handle many of the financial aspects of the cruise. For example, a purser handles the crew's payroll. This person also serves as the manager of the cruise personnel. The cruise director reports to the chief purser. Extensive leadership ability and financial experience are prerequisites for this position.

Numerous other cruise ship careers exist:

dining room captains	athletic/sports directors
dining room servers	social directors
stewards	youth activities coordinators
dancers	assistant cruise directors
musicians	casino workers
chefs	medical doctors
nurses	masseuses/masseurs
lecturers	

If this sounds like a lot of people work aboard a ship, you're correct. Typical is *The Song of America,* which sails out of the Port of Miami with a crew of more than five hundred!

GETTING STARTED

If a career on the high seas sounds a little too exciting, consider working for a passenger ship that cruises the inland waters or the waters along the coast. A number of opportunities exist in this area. The *Scotia Prince* operates out of Portland, Maine, and goes to the Nova Scotia area of Canada; a number of ships traverse the Atlantic intracoastal waterway—the *Newport Clipper* leaves Annapolis, Maryland, for Savannah, Georgia, at some times during the year and in cooler weather between ports further south on the coast—Palm Beach and St. Augustine, Florida. The *America* runs from Hilton Head, South Carolina, through Lake Okeechobee, Florida, to Fort Myers, Florida. Other ships follow a similar pattern, traveling to northern ports in warmer weather and southern ports in the winter. Likewise, this is true of the West Coast, where ships like the *Royal Princess* maintain an itinerary out of San Francisco, California, for Alaska and then return to Vancouver during the warmer months. As with most ships, which must remain active and income-producing, in winter the *Royal Princess* sails in southern climates.

A number of inland river cruises exist. The beautiful and historical *Delta Queen* and its modern sister ship the *Mississippi Queen* regularly paddle the Mississippi. The *Glacier Bay Explorer* provides inland water cruising on the West Coast in the Sacramento to San Francisco area. Further north the *Pacific Northwest Explorer* regularly leaves Portland in the spring and fall for a tour of inland waters.

If long periods of time away from home are a concern, there are many cruises that run for only a few days. Some are just three days, such as the *Canadian Princess,* which runs from Kingston to Brockville in Canada. The Hudson River Day Line has a one-day cruise from New York City to West Point Military Academy and back.

Most large ports have undergone a major renaissance, and almost all offer short cruises—usually for a couple of hours—to tour the port area. For example, *The Spirit of Philadelphia* operates out of Penn's Landing in Philadelphia with a number of daily cruises; *The Spirit of New Jersey,* which calls Weehawken, New Jersey, its home port, sails the Hudson River daily; the Circle Line regularly takes visitors on a water tour around New York City; and Miami's *Dixie Belle,* a paddle wheeler, has patrolled the inland waters with three cruises daily since 1950! These offer some opportunities for part-time and summer jobs for persons wishing to explore a cruise career afloat. Not only will you get the flavor of shipboard life, but you will also gain insight into some of the careers available. However, there are major differences between ship activities in the harbor and those at sea; a visit to a cruise ship when in port will quickly make these evident.

A word of caution about cruise ship careers: Some ships hire mainly employees from the nation of the ship's registry. For example, the *Song of Norway* (Port of Miami) has primarily a Norwegian crew, but it does employ citizens of the United States and Canada.

SHIP LINES

Those seriously interested in cruise ship careers may wish to purchase *Worldwide Cruise and Shipline Guide,* a bimonthly publication. Contact:

Official Airline Guides
2000 Clearwater Drive
Oak Brook, IL 60521

The following are cruise ship lines you may wish to contact if you have inquiries:

American Cruise Ship Lines, Inc.
One Marine Park
Haddam, CT 06438

American Hawaii Cruises
550 Kearny Street
San Francisco, CA 94108

Bahama Cruise Line, Inc.
200 Madison Avenue, Suite 500
New York, NY 10016

Bahama Cruise Line, Inc.
4600 West Kennedy Boulevard,
Suite 201
Tampa, FL 33609

Carnival Cruise Lines, Inc.
5225 Northwest Eighty-seventh
Avenue
Miami, FL 33166

Chandris Cruise Lines
Chandris Fantasy Cruises
900 Third Avenue
New York, NY 10022

Clipper Cruise Line
7711 Bonhomme Avenue
St. Louis, MO 63105

Commodore Cruise Line, Ltd.
1007 North American Way
Miami, FL 33132

Costa Cruises
  One Biscayne Tower, Suite
    3190
  Miami, FL 33131

Cunard Line, Ltd.
  555 Fifth Avenue
  New York, NY 10017

Delta Queen Steamboat Company
  30 Robin Street Wharf
  New Orleans, LA 70130

Eastern Cruise Lines
  1220 Biscayne Boulevard
  P.O. Box 010882
  Miami, FL 33101

Epirotiki Lines, Inc.
  551 Fifth Avenue, Suite 605
  New York, NY 10176

Exploration Cruise Lines
  1500 Metropolitan Park
    Building
  Seattle, WA 98101

Great Pacific Cruise Lines
  3600 Fifteenth Avenue West,
    Suite A201
  Seattle, WA 98119

Hellenic Mediterranean Lines
    Company
  One Hallidie Plaza, Suite 405
  San Francisco, CA 94102

Holland America Line-Westours,
    Inc.
  300 Elliott Avenue West
  Seattle, WA 98119

Home Line Cruises, Inc.
  One World Trade Center, Suite
    3969
  New York, NY 10048

International Cruise Center, Inc.
  185 Willis Avenue
  Mineola, NY 11501

K Lines-Hellenic Cruises
  Olympic Tower, Suite 504
  645 Fifth Avenue
  New York, NY 10022

Norwegian Caribbean Lines
  One Biscayne Tower
  Miami, FL 33131

Ocean Cruise Lines
  1510 Southeast Seventeenth
    Street
  Ft. Lauderdale, FL 33316

Pearl Cruises of Scandinavia
  1700 Montgomery Street, Suite
    320
  San Francisco, CA 94111

Premier Cruise Lines
  101 George King Boulevard
  P.O. Box 573
  Cape Canaveral, FL 32920

Princess Cruises
  2029 Century Park East
  Los Angeles, CA 90067

Regency Cruises
  260 Madison Avenue
  New York, NY 10016

Rhine Cruise Agency
  (Representing German Rhine
    Line)
  170 Hamilton Avenue
  White Plains, NY 10601

Royal Caribbean Cruise Line
  903 South America Way
  Miami, FL 33132

Royal Cruise Line, Ltd.
  One Maritime Plaza, Suite 600
  San Francisco, CA 94111

Royal Viking Line, Inc.
  One Embarcadero Center
  San Francisco, CA 94111

Salen Lindblad Cruising, Inc.
  133 East Fifty-fifth Street
  New York, NY 10022

Sitmar Cruises
  10100 Santa Monica Boulevard,
    Suite 1100
  Los Angeles, CA 90067

Society Expeditions
  3131 Elliott Avenue, Suite 700
  Seattle, WA 98121

Sun Line Cruises
  One Rockefeller Plaza, Suite 315
  New York, NY 10020

Sundance Cruises, Inc.
  520 Pike Street, Suite 2200
  Seattle, WA 98101

Western Cruise Lines
  150 West Sixth Street
  San Pedro, CA 90731

Windstar Sail Cruises
  7415 Northwest Nineteenth
    Street
  Miami, FL 33126

World Explorer Cruises
  550 Kearny Street
  San Francisco, CA 94108

Be alert to changes in cruise lines involving mergers, new organizations, and termination of service. The following organization offers updated information on its members' travel services:

Cruise Lines International
    Association (CLIA)
  500 Fifth Avenue, Suite 1407
  New York, NY 10110
  www.cruising.org/index2.htm

Two web sites exist that offer materials for prospective cruise ship employees.

The Real Truth About Cruise Ship Jobs: www.shipjobs.com/
Cruise Line Jobs: www.cruiselinejobs.com/

CHAPTER 3

OCEANOGRAPHY AND THE MARINE SCIENCES

Although many have a strong interest in oceanography, few realize what the career involves and how difficult it is for individuals to enter the exotic professions associated with it. Often *marine technicians,* persons who assist oceanographers, need to have a master's degree. Most oceanographers have a doctoral degree.

TYPICAL CAREERS

Oceanography involves the utilization of many sciences applied to the study of the oceans. It attempts to explain processes taking place in the ocean through the principles of other disciplines, such as biology, cartography, chemistry, geography, geology, medicine, meteorology, physics, mathematics, zoology, and others. Basically, oceanographers are professionals who use the sea as their laboratory for scientific study. Some rarely leave the laboratory; others do fieldwork in waters all over the world. Many serve on the faculties of universities, combining teaching and research. As with all professionals, each has a special interest within her or his chosen field. Although some overlap between specialties does exist, the main divisions are usually as follows.

Atmospheric Oceanographers

These scientists, also called *marine meteorologists,* study the relationship between the ocean and the atmosphere. For example, their work might involve analyzing the behavior of unusual weather such as thunderstorms, hurricanes, and waterspouts. Most recently, they have studied the El Niño phenomenon and its effects.

Biological Oceanographers

This field, probably the best known to the public, deals with the study of plant and animal life in the ocean. The interests of these professionals range from the one-celled phytoplankton to the largest of all life-forms on earth, whales. For example, some seek antibiotics from bacteria or fungi, others study new life forms found around hydrothermal vents such as giant clams and tubeworms. The improvement of sport and commercial deepwater fishing is also a concern of biological oceanographers.

Chemical Oceanographers

These professionals seek to understand the chemical reactions taking place in the sea. Also known as *marine geochemists,* they may study the food chain using radioactive substances. Often they will analyze the composition of pollution to suggest methods of prevention and removal. A growing concern with cleaning up the world's waters makes this field attractive to many.

Geological Oceanographers

These individuals, also known as *marine geologists,* investigate the ocean's floor and the shoreline, studying sediments, mineral content, and oil location. Recent interest in coastal erosion and in extricating minerals from the sea and oil from beneath the ocean floor makes employment opportunities good in this field. Some marine geologists obtain jobs with oil companies.

Physical Oceanographers

Persons in this field investigate physical properties of the ocean, such as the chemical properties of seawater, currents, tides, waves, and the relationship between the ocean and the atmosphere. Some study exotic topics such as the transmission of light and sound in the sea and the rise of ocean levels.

Oceanographic Engineers

These individuals design and build systems used in marine research and operations. They supervise the construction of underwater devices and floating structures such as oil drilling rigs and study the interaction of the sea with such structures.

Marine Policy and Law

The field covers the broad spectrum of ocean resource management issues concerning ocean environmental quality. Coastal zone management, fisheries, and marine archaeology are topics of interest to those working in this field.

As the work of oceanographers varies greatly, so do the working conditions. Those employed in laboratories examine organisms, conduct experiments, and analyze data. Others prefer to work in the field (or, in this case, the water). They utilize research vessels to aid them in their work. These men and women obtain samples from the sea and often must don SCUBA suits to collect specimens from the deep. Tidal pools, inlets, bays, and rivers also function as natural laboratories.

College and university instructors conduct experiments both in the laboratory and in the sea. Some oceanographers work in administrative and supervisory capacities. They manage research institutes, laboratory schools of marine science, and new product development in industry.

Closely related to oceanographers are *marine biologists*. The latter utilize the systematic procedures of the science of biology to study marine organisms. In the words of Professor Larry Harris of the University of New Hampshire, "Marine biology is biology with salt added."

REQUIREMENTS

A bachelor's degree is the minimum requirement for most persons wishing to enter the field. Consequently, a strong background in high school science and mathematics is particularly important. Many high schools now offer marine sciences courses that not only offer career awareness experiences but also provide a basis for future studies in oceanography. The University Laboratory High School at the University of Hawaii schedules a year-long marine science course for its students. It has become a model for many schools. You may want to suggest that your science teacher write for information:

Hawaii Marine Science Studies Project
 University Laboratory School
 1776 University Avenue
 Honolulu, HI 96822

Another school with a wonderful marine science program is located in Brooklyn, New York:

Fort Hamilton High School
 8301 Shore Road
 Brooklyn, NY 11209

Oceanographic technicians normally receive their training at a two-year community college or at a trade or technical school such as Leeward Community College (Hawaii), Florida Keys Community College, and the Southern Maine Vocational Technical Institute. Technicians assist in a variety of chemical and physical tests on waves, current flow, or water analysis. They maintain supplies and equipment in laboratories on shore and on research vessels at sea. Typical of their programs is the program in applied marine biology and oceanography at Southern Maine Vocational Technical Institute, which carries the following description:

> The program prepares students for jobs as research assistants or technicians. The curriculum emphasizes laboratory and field procedures commonly used by private, state, and federal agencies involved in marine and environmental quality research and monitoring.

The program, which leads to an associate degree, involves various oceanography experiences, including time aboard the school's ship, the *Aqua Lab III*. Some four-year undergraduate schools have a complete program in marine sciences. Many offer courses; a few provide for a major in oceanography. Students typically take courses like one offered at Oregon State University. Entitled The Sea, the course carries the following description:

> The basic principles of oceanography as demonstrated by current active research. Topics covered are: the formation of ocean basins, seawater, the atmosphere, and the origin of life; the interactions of the oceans, atmosphere, and climate; the geology of life of sub-sea volcanoes; waves, tides, and phenomena such as El Niño....

This is clearly an exciting and interesting course.

At Moody College of Marine Sciences at Texas A & M University in Galveston, Texas, they have initiated a new program entitled, Marine Sciences with a License Option: "The program retains the basic physical core of the marine sciences program but leads as well toward a United States Coast Guard License." A person who completes the curriculum can take the United States Coast Guard Examination as a third mate aboard a ship of any gross tonnage upon the oceans.

Combining course work in two areas assists the graduate in obtaining a job. Some colleges permit undergraduates to take courses during the summer at their marine laboratories. Some of these schools are Duke University in North Carolina; the Marine Biological Laboratory at Woods Hole, Massachusetts; the University of Texas; Drake University in cooperation with Southampton College (Long Island, New York); Stanford University in California; and the University of Washington.

On the West Coast, the University of Southern California (USC) has a fine facility—the Catalina Marine Science Center. Located on Santa Catalina Island, twenty-two miles from Los Angeles, the center engages in a wide variety of activities, including undergraduate and graduate teaching and research. The center sports some of the most innovative and contemporary oceanographic equipment.

It should be noted that some marine biologists strongly recommend that undergraduate college students major in biology with an occasional elective in marine biology, and then major in the marine sciences in graduate school. One of USC's master's programs is unique: the Master of Marine Affairs. The program contains a number of options but largely focuses upon marine policy. It provides sound training in management and basic work in marine sciences combined with three areas of emphasis—marine policy, marine law, and marine economics.

Most oceanographers pursue advanced study beyond their college bachelor's degrees. Scripps Institution of Oceanography of the University of California at San Diego suggests that undergraduate work include the following:

- mathematics—through integral and differential calculus
- physics—with laboratory, one year, using calculus in its exposition
- chemistry—with laboratory, one year
- an additional year in either physics or chemistry
- biology and geology—minimum of one semester each
- foreign language—preparation in at least one language (German, Russian, or a Romance language)

Long recognized for its leadership in the field, the University of Miami offers many graduate programs in the marine sciences. The study of sedimentation is typical of courses for those majoring in marine geology and geophysics, which carries the following description: "Environmental processes, and agents that form, transport, and deposit sediments, and the interpretation of the resulting rocks." Located in an ideal setting, the school's facilities encompass several laboratories and research vessels.

For more information relative to the oceanography/marine science field, some helpful web sites are:

International Oceanographic Foundation (at the University of Miami):
    www.rsmas.miami.edu/iof

Careers in Marine Biology and Oceanography:
    www-marine.stanford.edu/HMSweb/Career_booklet.html

Questions About Careers in Oceanography:
    www-ocean.tamu.edu/Careers/careers.html

The National Association of Laboratories:
    www.mbl.edu/html/NAML/NAML.html

The very helpful booklet *Marine Science Careers* is available from:

Communications Office
  Sea Grant
  Kingman Farm
  University of New Hampshire
  Durham, NH 03824-3512

An excellent resource is *University Curricula in the Marine Sciences and Related Fields,* which contains all college programs in marine science and lists the courses in each major field. It is available free by writing the following:

United States Department of Commerce
  National Technical Information Service
  5285 Port Royal Road
  Springfield, VA 22161

JOB OUTLOOK

The number of oceanographers remains small (five to six thousand); however, many persons are employed in related careers. The profession will undoubtedly grow in the future, particularly with expected expansion in private industry. The increased interest in and concern over the oceans will stimulate state and federal governments to increase employment in this field. Presently a large percentage of marine scientists (approximately 40 percent) work for the government. One such opportunity is with the National Oceanic and Atmospheric Administration (NOAA) of the Department of Commerce. NOAA oceanographers work in several exciting areas: some survey and chart the ocean floor, while others study the marine environment and its effect on the distribution and abundance of commercially available fish. Naturally, activities vary with the geography; oceanographers study the species peculiar to the area—salmon or tuna on the West Coast and lobsters or clams on the East Coast. For additional information contact the following:

NOAA
    Department of Commerce
    Personnel Division
    Operations Branch, AD41
    Rockville, MD 20852

Many other federal government agencies employ oceanographers. Write to:

Office of Personnel Management
    Washington Area Office
    1900 E Street NW
    Washington, DC 20415

Another large group of oceanographers/marine scientists is employed in academic and research positions in universities and marine science centers. The remainder work in private industry and consulting. Salaries are good to excellent, with ample opportunity for advancement.

SEA GRANT PROGRAMS

The Sea Grant Programs can provide additional information:

Alaska Sea Grant
    University of Alaska
    P.O. Box 755040
    Fairbanks, AK 99775-5040

California Sea Grant
    University of California-San
        Diego
    9500 Gilman Drive
    La Jolla, CA 92093-0232

Connecticut Sea Grant
    University of Connecticut
    1084 Shennecossett Road
    Groton, CT 06340

Delaware Sea Grant
    University of Delaware
    Robinson Hall, Room 111
    Newark, DE 19716

Florida Sea Grant
    University of Florida
    Box 110400
    Gainesville, FL 32611

Georgia Sea Grant
    University of Georgia
    Room 13, Ecology Building
    Athens, GA 30602-2206

Hawaii Sea Grant
University of Hawaii
1000 Pope Road, Room 223
Honolulu, HI 96822

Illinois-Indiana Sea Grant
University of Illinois
51 Mumford Hall
1301 West Gregory Drive
Urbana, IL 61801

Illinois-Indiana Sea Grant
Purdue University
Department of Forestry and
Natural Resources
1159 Forestry Building
West Lafayette, IN 47907-1159

Louisiana Sea Grant
Louisiana State University
128 Wetland Resources
Baton Rouge, LA 70803-7507

Maine/New Hampshire
Sea Grant
University of Maine
5715 Coburn Hall, Room 21
Orono, ME 04469-5715

Maine/New Hampshire
Sea Grant
University of New Hampshire
Kingman Farm
Durham, NH 03824-3512

Maryland Sea Grant
University of Maryland
0112 Skinner Hall
College Park, MD 20742

Michigan Sea Grant
University of Michigan
2200 Bonisteel Boulevard
Ann Arbor, MI 48109-2099

Minnesota Sea Grant
University of Minnesota,
Duluth
2305 East Fifth Street
Duluth, MN 55812

Mississippi-Alabama Sea Grant
Consortium
P.O. Box 7000
703 East Beach Drive
Ocean Springs, MS 39564

MIT Sea Grant
Massachusetts Institute of
Technology
Building E38, Room 300
77 Massachusetts Avenue
Cambridge, MA 02139

New Jersey Sea Grant
NJ Marine Science
Consortium
Building No. 22
Ft. Hancock, NJ 07732

New York Sea Grant Institute
State University of
New York
115 Nassau Hall
Stony Brook, NY 11794-5001

North Carolina Sea Grant
Box 8605
North Carolina State
University
Raleigh, NC 27695-8405

Ohio Sea Grant
Ohio State University
1541 Research Center
1314 Kinnear Road
Columbus, OH 43212

Oregon Sea Grant
Oregon State University
Administrative Services,
A500 G
Corvallis, OR 97331-2131

Puerto Rico Sea Grant
University of Puerto Rico
P.O. Box 5000
Mayaguez, PR 00681-5000

Rhode Island Sea Grant
University of Rhode Island
Narragansett Bay Campus
Narragansett, RI 02882-1197

South Carolina Sea Grant
Consortium
287 Meeting Street
Charleston, SC 29401

Southern California Sea Grant
Hancock Institute for Marine
Studies
University Park
Los Angeles, CA 90089-0373

Texas Sea Grant
Texas A & M University
1716 Briarcrest Drive, Suite 603
Bryan, TX 77802

Virginia Sea Grant
Virginia Graduate Marine
Science Consortium
Madison House
170 Rugby Road
Charlottesville, VA 22903

Washington Sea Grant
University of Washington
HG-30
3716 Brooklyn Avenue,
NE
Seattle, WA 98105-6716

Wisconsin Sea Grant
University of Wisconsin-
Madison
1800 University Avenue
Madison, WI 53705-4094

WHOI Sea Grant
Woods Hole Oceanographic
Institution
193 Oyster Pond Road
CRL 209
Woods Hole, MA 02543-1525

If you think marine science may be in your future, in addition to the suggestions contained in this chapter, read periodicals on the ocean (see Appendix A) and most importantly involve yourself in some pre-career experiences such as those mentioned in Chapter 10.

CHAPTER 4

FISHING

Often regarded as one of the oldest and most honorable of all professions, commercial fishing in recent years frequently has found itself at the center of controversies. The latter have ranged from fishing wars (over territory) to accusations of overfishing and disputes over management of catches to environmentally destructive use of drift nets ("walls of death") entangling large numbers of nontargeted fish. On a more positive note, these problems have focused attention on the importance of commercial fishing to regional and national economies and thereby provided support for the individuals and organizations involved.

This occupation has imprinted its mark on the North American landscape from the enchanting fishing villages of the Canadian maritimes and New England, to the shipjacks on the Chesapeake Bay, to the large tuna ships docked in the San Diego harbor. The state of Oregon alone licenses more than six thousand people to fish commercially!

Fishing has long been a family business in which the trade and its secrets are handed down from parent to child. Entrance to these jobs on small commercial fishing boats is difficult, but it is sometimes possible to sign aboard, expand your knowledge, and perhaps purchase your own boat or fleet!

Michael Bry lives in Ocean View, New Jersey, not far from the Atlantic Ocean. Upon graduation from Upper Merion High School in Pennsylvania, he moved to the South Jersey shore to pursue his career in commercial fishing. He had begun as a mate on a party boat during high school summers. Mike's first full-time job was as a crew member on an offshore tile and swordfish boat. This included weeklong trips to the Hudson Canyon, some ninety miles off shore. After two years, Mike looked for inshore work. He acted as a crew member on a lobster boat, also fishing for sea bass for one season. For a couple of years, Mike captained a gill-net boat, fishing for bluefish, shad, and weakfish.

Finally, Mike purchased his own gill-net boat, the *Andrea-M,* a forty-two-foot Virginia-style work boat Mike will operate out of Sea Isle City, New Jersey. Trading as the A & M Fish Company, Mike will wholesale fish to the Fulton Fish Market in New York City, where brokers will distribute it nationwide.

When asked what an average day is like, Mike replied, "long." It begins at 4:00 A.M. with coffee and the weather report. If all is as it should be, the boat is iced up for the catch and starts down the creek in the dark, with the help of radar. By daybreak, Mike and one other crew member reach the fishing grounds about six miles away, with color scope on and gill net ready for searching for fish. Mike reads the color scope, where fish appear on the screen in different shapes and colors. A buoy is thrown and circled by twenty-five hundred feet of net; then the boat is driven into the center and the fish swim into the net (thus, the name gill net). Hydraulics haul the net back into the boat. The catch is hand picked and put into holding tanks of saltwater and ice to keep it fresh and firm. Occasionally the catch is cleaned and gutted on the trip home to bring a higher price. At the dock, the fish are packed into cartons with ice and loaded on trucks bound for New York. Hopefully, the crew is home in time for dinner.

Salaries in the commercial fishing industry are dependent upon the type of fishing you do. Offshore fishing includes draggers, long-liners, and scallopers. A deckhand on any one of these boats makes an average of $50,000 a year. They work weeklong trips with little break in between almost twelve months a year! Those on factory ships—ships that both catch and process fish—will stay at sea until they catch their limit (all their ship can hold).

Clammers, gillnetters, and lobsterers make up the inshore fishermen. These people make daylong trips, dawn to dusk, nine months of the year on the East Coast, usually March through December. A crew member on an inshore boat makes approximately $15,000 to $20,000 per year. All commercial fishermen are paid a percentage of their catch. They are paid on a scale, the highest percentage to the owner, the next to the captain, first-mate, and on down. Offshore fishing is rated one of the most dangerous occupations; inshore fishing is considered safer.

Mike loves fishing for a lot of reasons; it's one occupation where the harder you work, the more you can make. A hard worker can make a good living, and as the owner of the boat, you can really push for success. "As an outdoor person, life on the ocean is really being away from the crowd." Mike can't see himself doing anything but fishing. When asked if he would recommend fishing to others he said, "If you love to fish and you're a hard worker, you can't beat it."

There is currently a strong movement among *marine economists* to help fishermen become good business managers as well as good fishermen. In the words of D. R. Hetchell, editor of *The National Fisherman:* "Unfortunately not every fisherman has the business training needed to operate independently in this modern world." Marine economists conduct workshops for fishermen, providing them with business knowledge and skills. Those considering a commercial fishing career should take business courses in high school or a community college.

Recently fishing has become more a science than an art. Boats use technical equipment to locate schools of fish. Fishermen know far more than their ancestors did about fish behavior, and catching strategies have improved. Much of this technological innovation is a result of a whole new breed of professionals known as *fisheries scientists*. Though not fishermen, these individuals contribute to the welfare of the industry. There exist a number of occupations in the area, and most fall into the following categories.

EDUCATORS

Persons working in this area fall into two camps: those who serve in universities and teach courses in fisheries science, and those who educate the general public about the sea. The latter are called *marine education specialists*. They work for aquaria, colleges, and marine institutes and frequently prepare publications for the media. Many teach programs about the sea for school-teachers who wish to integrate such information into their own courses. *Marine education specialists* work with fishing information, oceanography, and coastal zone management. This occupation's professional organization, The National Marine Educators Association, hosts a web site with excellent links at www.marine-ed.org/. Extension agents are educators who prepare reports, deliver talks, give advice to fishermen, and keep up with problems and trends in the industry.

ADMINISTRATORS

Administrators plan programs, coordinate budgets, and execute policies. They meet frequently with government officials (many fisheries administrators actually work in various government agencies) to discuss regulations and ways of improving business. A major responsibility of many administrators is public relations—working to improve fish sales and profit for the groups they serve.

RESEARCHERS

Areas of investigation for *marine researchers* include physiology, ecology, population dynamics, water quality, pathology, limnology, marine biology, and several other fields. While specializing in fish, these researchers must rely on related areas of knowledge. Usually, a researcher will deal with species from either freshwater or saltwater, but not from both. The former's work frequently relates to the stocking of bodies of water within a state. The researcher tries to maximize the success of the stocking program. Saltwater specialists work with all types of species but usually limit their interest to a few. Some engage in rather exciting activities, such as the study of shark behavior. Maybe you've read or seen on television the results of some of this work.

Often these scientists research mysteries. For example, phenomena that have occurred in Alaska among the Bering Sea fishermen have been the abrupt disappearance of their catch king crabs. This periodic appearance and disappearance has created havoc with people's lives, the economy, and the town of Dutch Harbor, Alaska. The latter has vacilated between boom, as North America's busiest fishing town, and bust!

Dr. Nancy Targett, a marine biologist at the University of Delaware, has researched artificial bait for the fishing industry. She hopes to find artificial horseshoe crab bait that works as well as the real one. Contributions such as this can preserve the horseshoe crab. See Chapter 3 for more on marine science careers.

MANAGEMENT POSITIONS

Managers work to maintain or develop fish populations for recreational or commercial reasons. Persons working in this capacity pay close attention to fish kills, monitor water pollution, estimate seasonal catches, reclaim new fishing grounds, work with conservation groups, and inform the public about the general state of fishing. One type of manager is the fish culturalist.

FISH CULTURALISTS

Fish culturalists, also known as aqua culturalists, manage public or private fish hatcheries, applying knowledge of management and fish culturing techniques. This management specialty requires a working knowledge of the nutritional and environmental needs of various species. General duties may involve overseeing, trapping, and spawning of fish; incubation of eggs; rearing of fry (young offspring); and the movement of fish to lakes, ponds, streams, or commercial tanks. A specific job might entail taking a species native to one area and introducing it in a new area. This is presently being done with Pacific salmon in Lake Michigan.

Persons in the fishery sciences have advanced training. Those who serve as assistants or technicians usually have at least an associate degree; many have a bachelor's. A typical two-year program resembles that of Ketchikan Community College in Ketchikan, Alaska. In addition to general courses, the following courses are taken in the fisheries science concentration:

fishing techniques	pot fishing
purse seining	refrigeration
gill net	hydraulic I and II
trolling	marine wiring I and II
fishing equipment repair	power mechanics
outboard engine repair	marine design
fish quality control	navigation and piloting
ground line	

Most of the others working in the profession have advanced degrees; some have doctorates, particularly those in research. In addition, federal government positions require a civil service examination. Approximately 135 colleges offer courses in fisheries biology; of these about 45 offer a major program leading to a degree. If such a program is consistent with your career goal, your high school courses should reflect college preparatory subjects with a science emphasis.

Typical of college level courses are the following, which form part of the program at the University of Minnesota:

Fishery Management. Fundamentals of population control; use of fishing regulations; habitat development; water quality control; use of artificial stocks for population maintenance; relationship between sport and commercial fisheries, including economic aspects; fundamentals of hatchery practice; and pond management.

Fishery Ecology in Polluted Waters. Description of degrading water quality control factors and influence on fish production. Fishery bioassay, setting of standards and determination of criteria for aquatic organisms, and administrative problems of pollution abatement. Biological effects of various pollutants on fish.

In Newfoundland, Canada, Memorial University houses the Fisheries and Marine Institute. One aspect of the Institute—the Fishing Technology Unit—works to improve the viability of fish harvesting through solving problems in the operation of fishing gear and equipment.

In addition to college and university degree programs, some institutions such as Clatsop Community College in Oregon offer certificate programs that focus on specific training for fishermen.

AQUACULTURE POSITIONS

This chapter will not be complete without a discussion of fish farming, also known as aquaculture or mariculture. This fast-growing industry involves growing fish in a systematic fashion in both saltwater and freshwater, indoors and outside. *Aquaculturists* conduct development and training programs for those in the business; they often write research proposals and consult with fish biologists and entrepreneurs to promote the industry. While a young industry, it has proven economically feasible and has the potential to provide quality protein. Undoubtedly it will continue to expand. Aquaculturists normally have a master's degree (some have doctorates); their educational background usually combines marine biology and business management. Opportunities and salaries appear very good; this truly will be a career of the twenty-first century.

JOB OUTLOOK

The employment outlook in fishing ranges from good to excellent, but it is sporadic. State and government positions will probably expand slowly, but qualified personnel are in short supply. The ranks of sports fishermen are expected to expand greatly; in the year 2000 an estimated sixty-three million anglers fished 1.3 billion days! This will obviously put a great strain on government fisheries agents and will require considerable expansion of their activities. Additional growth will occur in private industry, as we look to the sea to satisfy the nutritional requirements of the world's population. Off the East Coast alone it is estimated that there are twenty million tons of annual sustainable yield; about five million tons are presently taken.

The crafts and professions related to fishing offer a wide variety of career opportunities. If you enjoy life in coastal areas or on our inland freshwater systems and have an interest in fish, this career area deserves your serious consideration. One note of caution: Fishing careers are often demanding, requiring rigorous work under difficult conditions and in isolated locations. You may want to test your interest by visiting a hatchery or a marine science coastal laboratory, or try some involvement activities through the Boy Scouts or Girls Scouts or summer jobs.

SOURCES OF ADDITIONAL INFORMATION

Additional information also will prove helpful. Write the following:

American Fisheries Society
5410 Grosvenor Lane
Bethesda, MD 20014

American Institute of Fishery
Research Biologists
1226 Skyline Drive
Edmonds, WA 98020

American Institute of
Ichthyologists and
Herpetologists
National Museum of Natural
History
Washington, DC 20560

Environmental Protection Agency
Fourth and M Streets SW
Washington, DC 20460

Fish and Wildlife Service
U.S. Department of the Interior
Washington, DC 20240

National Fisheries Institute
1730 Pennsylvania Avenue NW
Washington, DC 20006

National Marine Fisheries Service
U.S. Department of Commerce
NOAA
Washington, DC 20235

National Wildlife Federation
1412 Sixteenth Street NW
Washington, DC 20236

Sport Fishing Institute
Suite 801
608 Thirteenth Street NW
Washington, DC 20005

U.S. Department of Agriculture
Fourteenth Street and Jefferson
Drive
Washington, DC 20250

Also write to the Department of Fisheries in your state; these will be located in the state capital. For information about opportunities outside the United States, write the following:

Fisheries Department
Food and Agricultural
Organization of the United
Nations
Via Delle Terme Di Caracella
Rome, Italy

Fisheries Research Board of
Canada
Department of the Environment
Ottawa, Ontario
Canada K1A 0H3

You may also wish to consult the Commercial Fishing Homepage web site:

http://sunsite.oit.unc.edu/london/...ulture/aquaculture///
msg00405.html

CHAPTER 5

COMMERCIAL DIVING

People's frenzy of interest concerning the world of water extends to the world beneath the surface. This exotic frontier of hot vents, unusual fish, and magnificent beauty has attracted the interest of people of all ages. Not all activities associated with the deep are exciting—some involve tedious manual labor related to commercial enterprises. In fact, the latter tends to be more common.

During his high school years, Dave Bollinger decided to enter the commercial fishing field. The opportunity for "travel, high pay, and adventure" attracted this experienced SCUBA diver to this romantic but dangerous field. Following graduation, he enrolled in the College of Oceaneering in Wilmington, California. The program included rigging, applications of mixed gas diving, and underwater welding, burning, and small tool usage. Since entering the field he has engaged in a variety of jobs: pipeline repair, contamination diving, cable laying, and others. The starting salaries are over $40,000 per year with extra pay for certain work; often the job involves overtime. Dave's advice to those considering this career: "The working conditions are hard; it's a dangerous job and it doesn't take much to die. It is something you should think about for a long time before getting into it."

The field of underwater diving is divided into four basic areas: 1. skin diving, in which you hold your breath while diving. In practice in North America, most skin divers use snorkels; 2. SCUBA (self-contained underwater breathing apparatus), in which you utilize a breathing device mounted on your back); 3. snorkeling, in which you use a breathing tube; and 4. surface air supply, in which air is pumped to you through a hose from the surface. Recently, a fifth category could be added—bell and underwater vehicle diving—in which you work below the surface inside a submersible.

While some forms of diving have been practiced for many years, it was not until the early 1960s that SCUBA diving moved from the status of a hobby to that of a profession.

SPECIFIC JOBS

Specific jobs in this industry vary greatly in terms of skills and work situations. The following is a sample list:

anchor and cable inspection
barge and pier construction and inspection
cable laying
coring of the ocean floor
demolition
dive resort activities
diving systems construction, repair, and maintenance
dredging
drilling equipment removal
emergency medical technician diver
equipment cleaning and installation
exploration drilling
fabrication of special equipment
flotation devices maintenance
flowline installation and maintenance
gas rack operation
gravity surveying
grouting
guideline replacement

harbor and deck construction
hyperbaric chamber operations
injection equipment installation
instrument use and maintenance
life support systems construction, operation, repair, and maintenance
marine environmental control work
marine patrol (law enforcement)
medical and emergency care for diving illnesses and accidents
oil drilling and oil production equipment installation, repair, and maintenance
operation of one-atmosphere suits, remote-control vehicles, and diving bells
pipeline installation and maintenance
platform construction
public safety and marine police activities
rock drilling and blasting
salvage
scientific and academic research activities:
    marine archaeology
    sample collecting
    seismic surveying
    specimen collecting
    surface geological appraisal
    surface mapping
search and recovery
sewage line installation, inspection, and maintenance
submersible operations
supervisor
tool design and testing
trenching
underwater diving instructor
underwater painting
underwater photography and videography

underwater welding and cutting
waterline installation, inspection, repair, and maintenance
well-head repair and maintenance

Many of these jobs require additional skills as well as good diving ability. Bob Smith, Director of the National YMCA Center for Underwater Activities, makes the point nicely: "Diving can be exploited to further another career." If you have the ability to weld, take photographs, or understand geology and you combine it with diving, your career can prove interesting and financially rewarding.

One career with considerable appeal is that of full-time diving instructor. This person may engage in full-time teaching, work in an underwater diving shop, or own or manage a school and conduct courses. In this managerial capacity, you might have the following program: courses on Monday nights at the community college; two courses on Tuesday and Thursday at the local high school; and private lessons at your school's pool.

Diving instructors often attend school. For example, those attending the NASDS (National Association of Skin Diving Schools) College receive the following training: diving physiology and psychology; prevention, diagnosis, and treatment of diving hazards; cardiopulmonary resuscitation techniques; certification in senior lifesaving; extensive work in surf rescue techniques and lifesaving adaptations to SCUBA; night diving; and deep diving and surf diving techniques. In addition, there are similar studies in teaching methodology, sales and marketing techniques, as well as general management. Linda Ponder, NASDS administrative assistant, reports, "Presently there are at least one hundred positions for qualified diving instructors open within our member SCUBA schools. They are in such demand that we cannot enroll enough students to fill the needs in our NASDS Diving Instructors College."

In some areas, SCUBA divers work on a contract basis with law enforcement officers. These divers are called upon when a person or a piece of evidence must be retrieved from the water. Several police departments have their own marine units that handle such matters. These officers may do regular police work and then swing into action as divers when the necessity arises. In some areas, such as Florida's Dade County, the Police Underwater Unit averages a call a day. The Florida Marine Patrol, a state group, has actually made some underwater arrests in such areas as illicit drug smuggling and lobster poaching! This career offers the opportunity to mesh an interest in police work with diving.

In recent years, an expansion of diving opportunities has resulted from the interest of the petroleum companies in offshore oil. This should continue strongly during the twenty-first century. These jobs involve divers with surface air supply or SCUBA. Surface air supply diving generally involves dives of great depths— one thousand feet is not unusual. A lifeline runs from the surface to the diver. SCUBA divers generally work at lesser depths and have the advantage of increased mobility. Some colleges, such as Santa Barbara City College, offer an associate degree (two years) in marine diving technology. Students take basic diving courses and technical subjects such as fundamentals of electronics and combination welding. Additional course work in oceanography, marine biology, and related general studies rounds out the program.

UNDERSEA SUBMERSIBLE DIVING

An area of diving that has received extensive media coverage recently is underwater salvage and treasure hunting. Perhaps you've read of projects to raise the sunken *Titanic* and *Andrea Doria* from their watery North Atlantic graves or to recover gold doubloons from Spanish galleys wrecked in the Caribbean. It should be noted, however, that few treasure-hunting projects ever raise the advance capital necessary for such a risky business, and of those that do get off the ground, only a few result in finds important enough to return the backers' investment.

Unromantically enough, in many cases the scrap value of a ship's hull and fittings may exceed any possible treasure value. Salvage divers find steady work more readily than their treasure-hunting counterparts. Nothing is wasted in the maritime field. When a ship is too old for commercial oceangoing use, it is often sold and refitted for inland waterways; twenty years later the same ship may be sold for scrap, even if it's lying beneath one thousand feet of water. Salvage divers with specialized training work at discovering and raising these ships.

Another underwater diving specialty receiving attention relates to research. That is utilizing underwater submersibles to obtain samples of marine organisms. Some individuals believe that the cure for many serious human diseases may be discovered in a marine plant or animal found deep in the ocean! Kim Reisenbichler serves as a senior research technician for Monterey Bay Aquarium Research Institute (MBARI). This organization, established in 1987, develops "state-of-the-art equipment, instrumentation, systems, and methods of scientific research in the deep waters of the ocean."

Kim credits his education for his position in such a valuable and interesting career. "I was introduced to oceanographic methodology while obtaining my A.A. degree at Fullerton College [California]. And I learned about biology, physiology, and some electronics while pursuing my B.A. and M.A. degrees at UC Santa Barbara." He strongly suggests, "For those who are contemplating entering this field, I cannot stress enough the need to gain practical experience." Kim has been a sport diving instructor and scientific diver for more than twenty years. He adds, "Good grades are essential—the thing that can set you apart from the pack is work experience along with good grades."

As one of those fortunate people who enjoys the beauty of nature under the sea while operating a remotely controlled vehicle and occasionally a manned submersible, Kim is the envy of many. He contributes to society through developing new sampling techniques for manned and remote vehicles.

California does not have a monopoly on the above activity; research activities take place in most waters of the globe. For example, Rutgers University operates LEO (Long-term Ecosystem Observatory) off the New Jersey coast. As part of this system, they operate an undersea robot! The latter travels in a preprogrammed area collecting data. Speculation exists that someday one of these vehicles might be utilized to provide data from Europa, one of the moons of Jupiter that appears to have an ocean beneath its icy crust.

The twenty-first century will demand a wide variety of specialists from scientists to technicians as we look for the oceans to provide a sea of knowledge benefitting people and society.

SCHOOLS

Two examples of schools offering programs for those interested in working with diving vehicles are the College of Oceaneering, and the Underwater Vehicle Training Center of the Professional School of Diving of New York. The former institution offers a course entitled, "Bell Diving." The program, which requires prerequisite diving course work and experience, covers "the disciplines of bell systems control; gas management; environmental control; communication; life support systems; decompression techniques; emergency procedures; and lock-out diving."

Located in Houston, Texas, the Underwater Vehicle Training Center offers a ROV (Remotely Operated Vehicle) operator/ technician course. The program of 320 hours spaced over eight weeks exposes the student to a wide variety of training including handling systems, vehicle propulsion, navigation and positioning, photography piloting, and leak detection.

A number of schools offer wreck diving specialty programs. The YMCA course covers topics such as the following:

ascending and exits
charter boat etiquette and safety
dive master duties
emergency procedures for handling decompression sickness
gearing up
introduction to wreck diving
life support equipment
marine life inhabitants in and around wrecks
motion sickness and diet
preservation of artifacts
rescue techniques and first aid
review of hazards of deep diving
state or federal laws on wreck salvage
wreck conservation and collection of artifacts
wreck-hunting techniques

Naturally, such a program requires good health and diving expertise as prerequisites to admission. In addition to classroom lectures, many of the courses involve dives at actual wreck sites.

Al Mikalow, director of the Coastal School of Deep Sea Diving in Oakland, California, reports that the commercial diving industry asks three things of a diver: Does the diver know how to use all kinds of equipment? Can the person handle many different kinds of commercial jobs? Can he or she perform these jobs under the most adverse conditions? At most schools the basic course is similar to that of the Divers Academy of the Eastern Seaboard in Camden, New Jersey. The sixteen-week program of 640 hours covers topics such as the following:

dive boat seamanship
diving equipment
diving physics and physiology
medical aspects of diving

SCUBA
techniques and procedures of deep-sea diving
underwater cutting and welding
underwater demolition

The Divers Institute of Technology located in Seattle, Washington, has a program that encompasses twenty-six weeks and 780 program hours. Its curriculum is more detailed than most and includes a segment on underwater television and video recording. Introductory courses are available in many locations around the nation. For information on advanced career-related programs, the following addresses will be helpful:

Coastal Schools of Deep Sea
   Diving
320 Twenty-ninth Avenue
Oakland, CA 94601

College of Oceaneering
272 South Fries Avenue
Wilmington, CA 90744

Divers Academy of the Eastern
   Seaboard
2500 Broadway
Camden, NJ 08104

Divers Institute of Technology
P.O. Box 70312
Seattle, WA 98107

Florida Institute of Technology
Melbourne, FL 32901

National Association of SCUBA
   Diving Schools
P.O. Box 17067
Long Beach, CA 90807

National Association of
   Underwater Instructors
22809 Barton Road
Colton, CA 92324

National Council of YMCAs
Diving Programs Department
291 Broadway
New York, NY 10007

Underwater Vehicle Training
   Center
10046 Chickasaw Lane
Houston, TX 77401

Some underwater careers are glamorous—a photographer who records the beauty of the undersea world is the envy of us all. Similarly, skippers of charter boats that take people on day trips and those involved in undersea habitat research can justly claim to lead interesting lives. Likewise, undersea treasure hunters may strike it rich as they prowl the ocean's depths hoping to locate wrecks of sunken ships. However, not all diving is as romantic as you may imagine. One of the reasons for the high salaries in this field is the hazardous nature of the work. Divers often work in cold, murky waters as well as in beautiful blue seas. Several factors influence the success of your work—depth, visibility, current, temperature, and the state of the surface. Another factor worth considering is that it takes three to four times longer to complete a task underwater than it does on land!

Persons wishing to enter this profession have done so through on-the-job experience, training while in the Navy, or formal training programs. However, in recent years formal training has become the principal vehicle for entrance into the profession. The future outlook of diving looks good, particularly for those with complementary skills in other areas. In fact, a worldwide deficit of trained underwater technicians has been predicted as exploration of the oceans expands. Diving employment is also steady with industrial companies (particularly the oil companies), government agencies, and small businesses.

One of the most glamorous and exciting water-related careers is commercial diving; the world beneath the sea offers awesome opportunities.

CHAPTER 6

MARITIME TRANSPORTATION

Andrea Weber Riniker serves as executive director of the Port of Tacoma, Washington. The latter, the sixth largest container port in the country and one of the largest in the world, handles more than thirteen million tons of cargo a year, largely from Asia. Although women remain underrepresented, in maritime transportation the situation is changing rapidly. Andrea believes, "These are great jobs, and I feel very lucky to have one. They're also jobs that I'd like to attract more women to. This is an increasingly global economy, and I want to see women comfortable in positions related to global trade and global economics."

North America's ports, open 365 days a year and 24 hours a day, form a vital component of maritime transportation. Port and harbor facilities provide employment for those engaged in the loading and unloading of commercial ships, in the inspection of cargo, in the underwriting of marine insurance and related capacities.

PORTSIDE OPPORTUNITIES

Stevedores (longshoremen) working in crews load and unload ships and move cargo in and out of warehouses. Much of the stevedore's work was previously simple human labor, but now some operate heavy equipment such as lift trucks and cranes.

The expansion of containerization has modified many of the stevedore's cargo operations. Some companies have cranes that can lift twenty- to forty-foot containers weighing up to forty tons and transport them along a 484-foot frontage area. The speed up and down the track is three hundred feet per minute. Despite this trend toward containerization, however, there are still many opportunities for longshoremen.

Wages vary with the kind of cargo handled; for example, the moving of distress cargo and explosives carries double the hourly rate of general cargo. Cargo that is thirty-two degrees Fahrenheit or below receive a slightly higher hourly rate than general cargo. Gang bosses supervise stevedores; that is, each stevedore is a member of a gang and operates as such.

Carloaders perform work related to the loading and unloading of railroad cars, trucks, barges, and containers.

Marine carpenters build grain feeders; lash and strap cargo using chain, wire, or rope; repair shipping pallets; and perform all wood-related work. They also crate and pack cargo.

Ship cleaners perform cleaning operations in the ship's hold. Specifically, they wash and paint surfaces, paint by hand and by spray gun, clean oil tanks, take inventory, clean and wash decks, clean and check lifeboats, clean quarters, and sort and check laundry.

Timekeepers govern the timekeeping of work performed on ships, piers, terminals, and barges.

In addition to the *mechanics, lockermen, gearmen, crane operators, truck drivers, guards, watchmen, ship maintenance personnel,* and others who work on the docks, several clerical occupations exist. *Billing clerks* type invoices that list the items shipped and the dates of shipment. *Manifest clerks* compile and type the list of the ship's cargo or passengers for use at custom houses and terminals. *Pilot station clerks* and *dispatchers* keep records of the ships entering and leaving harbors. *Checkers,* generally assigned at the rate of one per gang, note the receiving and delivery of all freight from vessel to pier or marine terminal. Salaries of dock and harbor workers are very good. For additional information contact the following:

International Longshoreman's
    Association
17 Battery Place, Room 1530
New York, NY 10004

National Maritime Union of
    America
346 West Seventeenth Street
New York, NY 10011

Steamship lines offer many opportunities ashore—the employment ranges from executive staff to office clerk similar to most corporations and embraces the following range of activities:

accounting
administrative
insurance
legal
marketing
operations
personnel
sales
traffic

Steamship companies also appoint agents in various cities to represent them and administer their activities. One could pick almost any major city and find opportunities in the steamship business.

One program that is unique exists at Texas A & M University in Galveston—the Bachelor of Science in Maritime Administration. Students take a curriculum composed of accounting, finance, law, and management combined with solid foundations in English, history, and liberal arts courses. In addition, course work in port operations, brokerage and chartering, maritime law, international trade and finance, and others round out this unique and exciting program.

Administrative and professional personnel work to bolster the success and efficiency of the port. They plan campaigns to attract ships to the port through personal contacts and brochures. For example, the South Jersey Port Corporation, an agency of the state of New Jersey, has considerably improved the tonnage of the port of greater Camden.

Each port has an organization (often called the Port Authority) that works for the benefit of shippers, unions, and the improvement of the economy of the port. These authorities also control the activities of the harbor. Several areas have marine trade associations that function as intermediaries in collective bargaining negotiations between shippers and labor unions.

The trade associations also plan and present training programs and perform safety inspections. The following association can provide the addresses of port authorities in your area:

The American Association of
    Port Authorities
   1010 Duke Street
   Alexandria, VA 22314

Also check the following web site with links to ports: http://marad.dot.gov/links/ports.html

Many harbor workers learn their craft on the job. Applicants for stevedore positions should be in good physical condition; with experience, persons with supervisory ability may become gang bosses. Some harbor workers become machinery operators and receive training from experienced equipment operators. Persons interested in technical or clerical positions should take appropriate course work in high school or community college. Previous work experience in an apprentice role also can be valuable. Many harbor administrators have college backgrounds; others have served as ships' officers or have come up through the ranks.

Employment opportunities will expand in the containerization field and are expected to decline in the traditional areas of cargo handling. However, Jim Traynor of the Philadelphia Marine Trade Association reports, "Pessimists continue to predict that containerization will have a profound effect on the number of available jobs, but all we can say is that it has not really happened in Philadelphia."

SHIPS AND SHIPPING

Moving commercial cargo by water is one of the oldest forms of trade. Raw materials, manufactured products, and liquids and solids travel by ships, barges, and boats. Each vessel must have a crew, whether it be a barge moving down the Allegheny River with coal, a towboat transporting grain along the Sacramento River, a supertanker moving in the Gulf of Mexico, or a container ship loading in Mobile, Alabama. Domestic shipping involves travel over the rivers, bays, lakes, and coastal waterways, while deep-sea ships traverse the sea lanes of the world.

Unfortunately the maritime shipping industry in the United States is experiencing serious problems. The number of United States flagships declines annually. This has had a serious negative effect upon maritime employment. Some seamen are able to obtain only four months of employment a year. On the Great Lakes only about one-half of the ships remain actively engaged in commerce. On the inland waterways there exist surplus barges. The expansion of container ships has greatly affected employment. For example, a large modern efficient container ship with a crew of twenty-one people can replace four smaller ships each with a crew of forty!

Naturally, shipping is affected by world economics; United States flagships, despite a magnificent record of reliability and safety, have difficulty competing with the ships of other nations. The latter pay their crews considerably less.

Although the United States fleet no longer occupies a preeminent position in world shipping, it maintains a position of leadership in maritime technology generally, and safety developments specifically.

Recently the world has seen the launching of such amazing new ships as supertankers and LNGs, liquefied natural gas carriers. The late Paul Hall, former president of the Seafarers International Union, explained the situation: "Our industry is in the midst of a major technological revolution...We are faced with the problem of crewing radically different ships like the LNG carriers—ships that require new and expanded job skills or proper manning."

Employment opportunities for young people without skills are rapidly disappearing; fortunately, several schools exist to train crew and officers for these new ships and for cruise ships.

The United States has been a commercial trading nation with a long and successful tradition. Even now that the ocean-shipping pace has slowed, our ships have an excellent record for reliability, safety, and environmental protection.

OCEANGOING OPPORTUNITIES

Historically, sailors have led rugged and interesting lives. Their careers took them to the ports of the world, and their work frequently tested their courage and strength. Today ordinary sailors still make up the majority of personnel aboard ships, which are the biggest self-sufficient mobile structures ever constructed. Most merchant seamen are assigned to a specific department and job title. This is a result of collective bargaining agreements between the ship's management and the unions. However, in recent years some personnel have received the rating of general vessel assistant, which allows them to do entry work in any of the three departments aboard ship.

The Deck Department

The beginning position in the deck department is that of *ordinary seaman*. This person functions as an apprentice working under the direction of able seamen in all phases of work. Some of the tasks performed are washing, painting, chipping rust and repainting, scrubbing decks, cleaning personnel quarters, and generally keeping the ship clean. The jobs will vary with the type of vessel. For example, on tankers the ordinary seaman also will be involved in cleaning the ship's tanks. The ordinary often will help others stand watch and will relieve the able seamen, who steer the vessel under supervision.

The mainstays of the deck force on all oceangoing vessels are the *able seamen.* For example, the *Queen Elizabeth II* employs thirty-two able seamen. Making up about one-fifth of all sailors, they must have a thorough knowledge of all parts of the ship and be able to use all gear and deck equipment. They are responsible for all of the deck rigging and machinery, cleaning, and painting. On some ships, they serve as *quartermasters* or *helmsmen* that steer the ship. They must steer the vessel according to the direct orders of the ship's officers. Officially, they must understand steering commands, elementary construction and parts of the ship, nautical terms, passing, fog and distress signals, running lights (lights carried on ships by international agreement between sunset and sunrise), the compass, and many other subjects.

In emergencies these seamen perform many important tasks. During a fire, they close doors and handle fire equipment. In serious difficulties, they assist with the lifeboats—removing covers, swinging the lifeboats free from the ships, and lowering them into the water. They must hold lifeboatmen's certificates and be expert in their use. Able seamen practice emergency procedures to keep their skill levels high. Upon docking they handle lines, assist with loading and discharge of cargo, and sometimes clean the cargo compartments.

The *boatswain* (or *bosun*) serves as the highest ranking seaman in the deck department. As foreman of the deck crew, the boatswain assigns work and supervises general maintenance operations. This person meets with the chief officer to plan the day's jobs, which may involve overhauling the lifeboats or cleaning and polishing the metalwork. Some vessels carry a *ship's carpenter* who works with other deck department personnel in making repairs. The title originates from the days of wooden ships and doesn't accurately describe the carpenter's real duties, which may involve replacing broken glass in a porthole, repairing a lock, or fixing a broken anchor. On some ships, *deck utility people* (*deck maintenance people* or *deck/machine people*) perform deck maintenance work under supervision of the boatswain.

The Engine Department

American ingenuity contributed to the development of the steamship and, consequently, to the demise of the sailing ship. First wood, then coal, and finally oil served to fuel these new vessels; a short experiment in commercial nuclear propulsion (the NS *Savannah,* named for the first steamship to traverse the Atlantic) proved too expensive. The move away from sails resulted in a new specialty aboard ship; those who tended the area of the ship responsible for propulsion. Today various specialists make up the staff that care for the engines.

The *wipers,* the lowest ranking in the department, perform general duties. These persons clean and help repair engine room equipment. *Oilers* lubricate the moving and wearing parts of the engine machinery. They check oil pressure and flow to prevent engine damage and aid in the overhaul and repair of engines. *Fire-watertenders* check and regulate the amount of water in the boilers, inspect gauges, and regulate fuel flow to keep steam pressure constant. The *qualified member of the engine department* (*QMED*) must be able to perform and assist with all unlicensed machinery work, such as the refrigeration and steering systems, and must know how to operate the new automated engine room consoles. Aboard tank vessels is the *pumpman,* who assists the cargo officer in the handling of liquid cargo and in related tasks such as maintaining pumps and valves. The generator and motors are serviced by the *ship's electrician.* This person, sometimes called the *chief electrician,* may have the assistance of a *second electrician.* The replacement of defective lights and fuses, along with other routine electrical repairs, falls within the domain of electricians.

The Steward's Department

Messmates and *utilitymates* form the lowest rating within this department. The former set the tables, serve meals, and clean the mess rooms. The latter prepare vegetables and clean cooking utensils and galley (kitchen) equipment. The *assistant cook,* working under the direction of the chief cook, prepares the food for cooking. Similarly, the *cook* and *baker,* under the leadership of the chief cook, do all baking and cook all vegetables and hot meals. In charge of the galley is the *chief cook.* The chief cook directs the preparation and serving of all meals; the orderliness and cleanliness of the galley also fall under this person's direction. The supervision of the department falls under the jurisdiction of the *chief steward.* Ordering supplies, planning menus, and supervising meals are some of his or her duties. The chief steward, in consultation with the master, authorizes overtime, receives stores, and accepts responsibility for the efficiency of the department. As with other ratings aboard ship, on many new ships some jobs are combined. For example, the chief cook and chief steward have been replaced by a combination chief steward/cook on some vessels.

EDUCATION AND TRAINING

Some high schools have begun to offer courses for those interested in maritime transportation careers. At this point in time, most tend to be marine science offerings; however, at least two schools offer excellent maritime studies programs. The Randal Aerospace and Marine Science High School in Washington, DC, gives courses in basic seamanship. These are taught in conjunction with the harbor police and include boating, navigation, "rules of the road," boat maintenance, and some maritime law. In the marine engineering course, students are introduced to the maintenance of marine engines, specifically the 671 GM Diesel Generator, while learning the use of basic hand tools and principles of preventive maintenance. Upon graduation, many of the students obtain jobs in marine and maritime industries. Dr. Richard Odgorny, program director, reports that graduates enter such careers as marina steward, fireboat firefighter, deckhand, the Navy, sailing instructor, lifeguard, crabbing and fishing apprentice, yacht basin service station attendant, marine engine mechanic's helper, boatbuilding, and clerical positions in government agencies. Their salaries were significantly higher than the average for all high school graduates. The student population is divided almost equally between men and women.

Several colleges now offer courses for prospective sailors, and the list is growing. Typical of these institutions is Leeward Community College in Hawaii, which provides such courses as advanced seamanship, basic seamanship and marine orientation, and practical shipboard operation. To locate colleges with maritime offerings in your area, work with your school counselor and use college guidebooks. Several programs exist in community colleges. The following is only a geographic sampling of the many programs across the nation:

Delgado Community College
   Fire School
   13200 Old Gentilly Road
   New Orleans, LA 70129

Leeward Community College
   96-050 Farrington Highway
   Pearl City, Oahu, HI 96782

Santa Barbara City College
   721 Cliff Drive
   Santa Barbara, CA 93109

Seattle Central Community
   College
   4455 Shilshole Avenue NW
   Seattle, WA 98107

Southern Maine Vocational
    Technical Institute
Fort Road
South Portland, ME 04106

Suffolk County Community
    College
Eastern Campus
Speonk Riverhead Road
Riverhead, NY 11901

The importance of professional training to members of the merchant marine is reflected in the comments of Frank Drozak, president of the Seafarers International Union:

> A near tragedy was avoided this month by the alertness of the SIU-contracted *Seattle,* and the excellent training in safety techniques many of its crew had received.
>
> An engine room fire broke out, November 7, while the *Seattle* was in the middle of the North Atlantic. The oiler on watch had been through fire-fighting school, so he knew just what to do when he saw a fire starting in the auxiliary generator. The day-working wiper, who had also been through fire-fighting school, helped him get the blaze under control...the ship was eventually saved, although the fire lasted six hours.

Fire is the greatest safety hazard aboard ship, and modern marine technology has resulted in new and different situations in which fires can break out. Mariners have limited options in the event of a fire—they cannot simply walk away from the flames to safety. The potential disaster to human life, the cargo, and the multimillion dollar ship is great. Training for prevention of fire-related tragedies is a must.

The preceding example is just one illustration of the importance of professional training for a maritime career. Few opportunities remain for the unskilled in today's merchant marine; fortunately, several schools and programs exist for this purpose. More important, every person employed aboard an oceangoing or coastal United States vessel of one hundred or more gross tons must obtain a United States merchant mariner's document issued by the Coast Guard. Application for the certificate must be endorsed by a recognized maritime school or by a letter of commitment from a shipping company. Today, few companies make such a commitment. Coast Guard or naval experience also meets this requirement.

One of the best known professional schools in the industry is the Seafarers Harry Lundeberg School of Seamanship (SHLSS). At this tuition-free school, training is provided in entry or beginning job skills, upgrading to licensed or unlicensed ratings, and specialized, advanced courses. The school focuses its program in three key areas:

1. providing young men and women who have no maritime experience with the basic skills they will need to serve aboard U.S. flagships or tugs and towboats,
2. providing professional advancement for experienced men and women through career upgrading programs, and
3. providing the academic education that is an essential complement to the modern technical skills in today's water transport industries.

Part of the curriculum involves training aboard ship. The program includes courses such as the following for each trainee:

Vessel familiarization. This course is designed to acquaint students with basic seamanship skills and shipboard terminology. Students learn marlinspike seamanship and cargo handling applicable to ocean and inland waters. Safety, watchstanding duties and work routines also are taught. On-the-job training takes place on the school's ships and tugs.

Lifeboat/water survival. This course of instruction leading to a lifeboatman certification consists of classroom work and practical training. Subjects taught include emergency drills, lifeboat launching and recovery, lifeboat construction, basic compass navigation, life raft construction, life raft launching and maintenance, and use of all lifeboat and life raft equipment.

The SHLSS offers numerous courses for seamen to upgrade and improve their skills. In the engine department, course work may be taken to obtain an endorsement as fireman/watertender. In the deck department, upgrading courses carry titles such as quartermaster, radar observer, and deep-sea deck shiphandling simulator. Cook and baker, chief cook, and chief steward are some of the advanced courses in the steward department.

For additional information, contact the school directly:

The Seafarers Harry Lundeberg
    School of Seamanship
    Piney Point, MD 20674

You can also contact the Seafarers International Union, which administers the school at www.seafarers.org/

Another fine training program is the National Maritime Union (NMU) Upgrading and Retraining Plan. The purpose of the program is "to provide experienced deck and engine department seamen with training in the new skills, knowledge, and techniques required on automated ships." The program also provides training in shipboard service and food preparation, as well as efficient management in the steward's department. The basic program is fifteen weeks long with specialization taking up to twenty-one weeks. The program is designed for NMU members, not for those just beginning a maritime career. Travel expenses, board, and a stipend are provided. For further information write the NMU:

NMU Upgrading and
    Retraining Plan
    346 West Seventeenth Street
    New York, NY 10011

Marine Labor Operated Schools

The following schools offer a variety of programs for those interested in a career afloat:

ARA TIME
Department of MITAGS
5700 Hammonds Ferry
Road
Linthicum Heights, MD 21090

Calhoun MEBA Engineering
School
Route 5, Box 509
Easton, MD 21601

District # MEBA—AMD Safety
and Education Plan
School of Marine Engineering
and Navigation
2 West Dixie Highway
Dania, FL 33004

Houston Marine Training
Services
1600 Twenty-ninth Street
Kenner, LA 70062

Maritime Administration
Eastern Region Fire Training
Registrar
26 Federal Plaza, 37th Floor
New York, NY 10278
(Each of the regional offices of the
maritime administration
offers fire-training courses.)

Maritime Institute of Technology
and Graduate Studies
5700 Hammonds Ferry Road
Linthicum Heights, MD 21090

Maritime Safety Consultant/
Tidewater School of
Navigation
Lonsdale Building—207
Granby Mall
Norfolk, VA 23510

Marine Safety International
Marine Air Terminal
La Guardia Airport
New York, NY 11371

MEBA—AMO Great Lakes
School
1 Maritime Plaza
Toledo, OH 43604

Radio Officers Union, District 3
NMEBA (AFL-CIO)
Maritime Electronic Training
School
Route 5, Box 509
Easton, MD 21601

Sea School
3770 Sixteenth Street N
St. Petersburg, FL 33704

A variety of other programs exist in the maritime field. The Franklin D. Roosevelt Institute of the Seamen's Church Institute of New York offers as one of its programs a Certificate Program in Maritime Transportation. The program involves six courses, including Modern Ocean Transportation, Maritime Law, and Container Control, and Terminal Operations. For information about this and other programs, contact the following:

Seamen's Church Institute
of New York
15 State Street
New York, NY 10004

Another organization you may wish to contact is

American Maritime Officers
2 West Dixie Highway
Dania, FL 33004

OCEANGOING WORKING CONDITIONS

Working conditions vary with the nature of the route traveled and the type of ship. The workweek is considerably different from that on land. At sea, most sailors are required to work seven days a week—generally in two separate four-hour shifts every twenty-four hours. While persons in the engine room must be able to withstand intense heat, deck workers have to contend with rain and cold. Modern ships have clean, comfortable quarters. On older ships, rooms are shared; on newer ones, each crew member has a room. Meals are taken in the mess hall, and the food is usually plentiful and good. American sailors have some of the highest salaries in the world; they also earn ample overtime. Some Americans work on the ships of other nations or on vessels owned by American companies and registered abroad. Fringe benefits such as insurance and vacation time are very liberal, and opportunities exist for advancement. Additionally, seafarers have the chance to travel the world. However, with the trend toward quick turnaround time in port, touring is limited.

If this lifestyle appeals to you, you can obtain additional information by writing to the various maritime schools and associations.

MERCHANT MARINE OFFICERS

Today's ships, costing in the hundreds of millions of dollars, require expert knowledge and skills on the part of the ships' officers. Modern vessels perform some amazing tasks—some barge ships can carry burges weighing five hundred tons each on three decks, and many container ships carry 2,100 forty-foot containers at average speeds of over twenty-five miles per hour.

In command of every oceangoing vessel is a *captain* or *master.* The captain has complete responsibility for the ship's operation and the safety of the crew, passengers, and cargo. While at sea, the captain has complete authority to run the ship within certain guidelines. In port, the captain may serve as the shipowner's agent in relations with port officials and may act as paymaster. Captains coordinate water and food supplies, piloting, towing, and stevedoring services.

The head of the deck department is the *chief mate* or *chief deck officer.* This person serves as the captain's assistant in assigning duties to the deck crew and in managing personnel. The chief mate, working with the deck crew, plans and supervises the loading and unloading of cargo and assists the captain and pilot in taking the ship in and out of port. To comply with Coast Guard regulations on the safe and efficient operation of ships, deck officers must be certified for radar, gyrocompass, and radio directional equipment.

The *second mate,* by tradition, serves as the navigation officer. This person maintains navigational equipment, checks instruments, and makes sure that appropriate, up-to-date charts are available for navigation.

Some *third mates* generally function as signal officers. They coordinate loading and unloading cargoes, inspecting lifeboats and lifesaving equipment, and determining the ship's position. Most deck officers serve watch.

The *chief engineer* supervises the engine department and its personnel and has responsibility for the safe and efficient operation of the engines and mechanical equipment. While at sea, this person oversees the operation of the propulsion system and auxiliary equipment and performs other functions such as calculating fuel consumption.

Serving under thc chief, the *first assistant engineer* supervises the engine room personnel. The maintenance of equipment and machinery falls under this person's direction. The boiler and related equipment come under the direction of the *second assistant engineer,* who ensures that proper steam pressure and oil and water temperatures are maintained. The *third assistant engineer* supervises the lubrication of the engine room equipment and distillation of saltwater to freshwater. On some ships, the electrical and refrigeration systems also fall into the domain of the third assistant engineer. Also serving aboard ship is the *radio officer,* who sends and receives messages from shore and from other ships. Passenger ships and some freighters may employ a *purser.* This person assists passengers, assists with the paperwork necessary for a ship to enter or leave port, and, in some cases, serves as the physician's assistant.

Certain requirements must be met to enter the ranks of officers. In addition to age and health qualifications, the candidate must have three years of work experience aboard ship or be a graduate of an approved training program. Coast Guard examinations test the knowledge of the candidates. To advance in rank, officers must pass progressively more difficult examinations. Although no formal education is required, most officers attend school. The surest way to become an officer is to attend one of the fine programs at the maritime academies.

Maritime academies across the country are committed to preserving and fostering our country's maritime tradition. Their impressive programs—theoretical knowledge combined with practical experience of life aboard ship—can be immensely rewarding and enjoyable. At the same time, maritime programs are not without challenge, as illustrated by the comments of Gregory Snyder, a former midshipman at the United States Merchant Marine Academy:

> The most difficult challenge in coming to Kings Point was adapting to a complete and radical change in lifestyle. I had to be willing to exert my greatest effort in everything I did. I feel that Kings Point has made me more responsible, more confident, and a much harder worker.

In addition to general studies courses, comparable to those at regular colleges and universities, the maritime academies require course work in maritime history, marine chemistry and engineering, oceanography, international trade, all forms of electronic navigation, and cargo operations. While most college students attend class for nine months each year, maritime academy midshipmen have eleven months of school. The longer program reflects the time necessary for students to obtain their practical training.

It should be noted that all maritime academies welcome qualified women candidates. In fact, the Texas Maritime Academy and the Maine Maritime Academy have the joint distinction of having graduated the first women in the history of the United States Merchant Marine. Deborah Doane from Maine and Susan Carter from Texas both graduated in 1976.

For further information on programs, prerequisites, and fees, contact the following:

California Maritime Academy
  Vallejo, CA 94591

Great Lakes Maritime Academy
  Northwestern Michigan College
  Traverse City, MI 49684

Maine Maritime Academy
  Castine, ME 04421

Massachusetts Maritime Academy
  Buzzards Bay, MA 02532

New York State Maritime College
  State University of New York
  Fort Schuyler, NY 10465

Texas Maritime Academy
  Texas A & M University
  Galveston, TX 77553

United States Merchant Marine
  Academy
  Kings Point, NY 11024

A superb web site entitled Maritime Education contains links to each of the maritime academies: http://ils.unc.edu/maritime/mared.html

The programs vary at the above institutions. For example, the Great Lakes Maritime Academy offers two three-year associate of applied science programs: one leads to a Coast Guard license as a Great Lakes mate, and the other leads to a Coast Guard license as a third assistant engineer. New York State's Maritime College offers the following academic programs: Marine and Power Systems Engineering, Marine Transportation/Business Administration, Marine Environmental Science, Naval Architecture, Facilities Engineering, and Humanities. The following is a sample of titles of specialty courses: Nautical Science—Tugs and Towing, Marine Transportation—Marine Insurance, Law—Admiralty Law.

In Canada, Dalhousie University, located in the port city of Halifax, Nova Scotia, offers a unique program entitled, Master of Marine Management. It features courses "on all aspects of the uses of the seas" such as Coastal Zone Management, Marine Science and Technology, and Contemporary Issues in Ocean Management and Development. For additional information consult: http://www2.dal.ca/mmm/

The future of maritime enterprise promises that those aboard ship or in harbor activities will lead exciting lives. Ship's officers and harbor supervisors will be in the forefront of many new developments. Salaries, working conditions, and vacations are very attractive. Job opportunities are fair, and during good times off shore oil drilling is excellent. Little-known opportunities exist for academy graduates among the Army Corps of Engineers' dredging operations, as pilots and tugboat captains in the Panama Canal Zone, and aboard research vessels belonging to the National Oceanic and Atmospheric Administration and the United States Navy and Coast Guard. Researchers have predicted a moderate increase in positions in most areas. If certain factors occur, there could be a serious shortage of maritime officers in the twenty-first century.

Those with an interest in the maritime industry and a desire to serve in a position of leadership should seriously consider a career as an officer.

INLAND MARITIME CAREERS

A recent issue of the *Shipjack*, the newspaper of the Seafarers Harry Lundeberg School of Seamanship, posed the following question: "What is a quarter mile long and carries as much as a thirty-five-mile convoy of trucks...or as much as twelve 100-ton jumbo railroad cars or as much as sixty 25-ton semitrailers?" The answer: A tow of fifteen barges! As this country has become more conscious of the need to conserve energy, water transportation continues to be recognized as extremely efficient. It is for this and other reasons that the domestic shipping industry should experience expansion.

Presently, some 15 percent of America's total transportation moves on its inland waters; 4 percent moves on the Great Lakes (Duluth is the nation's number-one port by tonnage); and 5 percent moves along the coastal ocean routes. During the early part of the twenty-first century, river traffic is expected to increase, Great Lakes traffic is expected to increase, and coastal waters traffic will experience significant growth.

Most freight on the inland waterways moves by unmanned, non-self-propelled barges having drafts of six to fourteen feet. The minimum depth of navigable waterway is nine feet harbor; channels are maintained at forty feet by the Army Corps of Engineers. Part of our inland commerce also takes place on shallow, self-propelled vessels and tankers. That is, either a tug pulls them or a towboat pushed them. Occasionally a tug will push the barges or carry them alongside.

Barges transport both dry and liquid cargo over the more than twenty-five thousand miles of America's inland ways. Barge sizes range from five hundred to three thousand tons; however, some oceangoing barges have a capacity of more than thirty thousand tons. The cargoes particularly appropriate to this mode of transportation are raw commodities: petroleum products, chemicals, grains, forest products, and iron and steel.

Increasing demand for water transportation has stimulated the improvement of the waterways, equipment, and technology. Consequently, barge service is safe and reliable. Origin-to-destination speed averages six miles per hour with some tows making fifteen miles per hour. Towboats and barges are under construction and skilled workers are needed to work on them.

The attraction of this profession can best be summarized in the words of Charlie Piot. A Princeton graduate, Charlie was drawn not by the financial opportunity of the river, but by his personal interest in it.

> I came to the river two years ago, just out of college. I wanted to wear a new shirt for awhile, to see things from a different angle. The pattern of my life has been all too much the same—my childhood in the suburbs, my high school years in a private school, my college years in an eastern university. It was all too narrow and too simple—a well-lubricated rail that, once hitched up to, I could quite easily follow for the rest of my earthly life and never see another side.
>
> So I sought out work on the boats. I often wondered what life was like on those sooty barge tows I'd seen creeping past the St. Louis waterfront. What was the work like? And the men who work with their hands and their bodies to make a living—what were they like? And the river—the one we've all vicariously experienced through Twain—what was its mystique?
>
> I petitioned the offices of several barge lines and landed a job as a deckhand on a towboat that runs the lower Mississippi from St. Louis to New Orleans.

Charlie found the work rigorous and life on the river rewarding; he expanded his original plan of working six months to two years.

Skill, ability, experience, endurance, and a feel for the river are the prime requirements necessary for success in towing.

Personnel

In the towing industry, the *deckhand* serves as a jack of all trades. This position varies widely with the nature and the size of the cargo and towboat. On large riverboats, the deckhand ties together the many barges into unit tows and breaks them apart upon reaching the destination. On harbor tugs, the responsibility for tying the large ships to the tug falls to this person. Deckhands usually work a schedule of six hours on, six hours off. Boatmen usually spend a specified number of days on and off their jobs each month; twenty days on, ten days off is one common arrangement. This type of work schedule combined with the rigorous and sometimes dangerous nature of the work constitute a lifestyle that is not for everyone.

One of the most important positions in the industry is that of *tankerman.* The movement of millions of barrels of petroleum and other liquid cargoes each year has made this a critical and rapidly expanding job. The tankerman has the responsibility for cleanly and safely loading and unloading the cargo. Once under way, the tankerman maintains close surveillance of the liquid, checking pumps and engines. Some persons in this position also work in harbors, refueling ocean ships from bunker barges. Knowledge of and adherence to environmental protection standards is important in this job. The handling of hazardous cargo also requires Coast Guard licensing. This regulation, now under study, will probably become more stringent.

In general, towboats have two operators—the *captain* or *master* and the *pilot* or *mate,* who serves as the captain's assistant. When under way, the operators each stand two six-hour watches per day. The responsibility for the safe navigation of the vessel falls to them. While on watch they steer the vessel, monitor the radio and radar sets, and fill out the log. The master works aboard oceangoing tugs; pilots often work on harbor tugs. Aboard towing vessels operating on longer routes, a captain, first mate, and second mate are required. They stand two four-hour watches per day. Boats working the western rivers and inland waterways of the Gulf of Mexico employ a *steersman.* This person, usually in training to become a mate or pilot, steers the vessel. On some of the larger boats an *engineer* maintains the engine room. Deckhands often perform this function on smaller boats. The *towboat cook* has the responsibility of serving and preparing all meals for the crew; this includes four meals a day and coffee around the clock. On small boats, deckhands take turns doing the cooking.

EDUCATION AND TRAINING

The Seafarers Harry Lundeberg School of Seamanship and a number of regional schools offer education and training programs. SHLSS Inland and Near Coastal programs have courses such as Rules of the Road, Shiphandling and Seamanship, Cargo Handling II and others.

Other regional schools also exist:

Louisiana Marine and
    Petroleum Institute
P.O. Box 10251
Station 1
Houma, LA 70363

You may also wish to write for information from the following:

The American Waterways
    Operators
1600 Wilson Boulevard,
    Suite 1000
Arlington, VA 22209

Continued slow expansion of the industry has created a concern that sufficient personnel will not be available to staff its expanding positions. In fact, *Waterways Journal,* the industry's magazine, has predicted an increase in personnel. Presently, more than fifty thousand people work aboard inland boats and barges, and a similar number find employment in shore activities related to inland shipping. The increasing cost and complexity of barges, tugs, and towboats make the acquisition of specific skills particularly important to persons interested in this career area.

A web site, entitled "Maritime Links," provides numerous excellent links: www.csum.edu/link/.

On a smaller scale a number of schools providing boat-related training exist; for some these are an entrance to this field. For example, the Annapolis Sailing School/Annapolis Powerboat School offers instruction to recreational boaters in Annapolis, Maryland; St. Croix, Virgin Islands; Marathon, Florida; and St. Petersburg, Florida. It is located at 601 Sixth Street, Annapolis, MD 21403.

In a different vein, the American Sailing Training Association located in Newport, Rhode Island, takes as its mission, "To provide people from all walks of life with the opportunity to experience seafaring aboard a vessel under sail." For programs they certify and sponsor contact: http://tallships.sailtraining.org/

As vessels vary greatly, so do careers afloat; all contribute to an enjoyable, rewarding lifestyle.

CHAPTER 7

SHIPBUILDING

The United States has had a rich and productive shipbuilding tradition that predates its nationhood and has enjoyed many notable successes. Very shortly after its independence, the U.S. ranked as a world leader in shipbuilding. Later, American steamships would capture world attention and, along with British developments, would sound the death knell of wind-powered sailing ships. On the inland waterways, paddle wheel steamboats reflected the charm and tradition of nineteenth-century rural America. Likewise, Canada has a strong shipbuilding tradition; the Versatile Pacific Shipyards in North Vancouver are but one example.

The development of the submarine represents a genuine American success story. Pioneered during the American Revolution, bolstered by Confederate technology during the Civil War, and presently exemplified by nuclear-powered denizens of the deep, the submarine reflects true American ingenuity.

Shipyards were again put to the test during World War II, when more than five thousand vessels were produced for the war effort. More recently, American shipyards have suffered from neglect. Many now emphasize repair as opposed to construction. Typical is Norfolk Shipbuilding and Drydock Corporation (NORSHIPCO), one of North America's leading ship repair complexes located at the port of Hampton Roads in Virginia.

Historically, ship construction has followed these steps:

1. *Preliminary design phase.* Today's craftsworkers rely heavily on computer aided design throughout the construction process.
2. *Contract award.* This marks the initiation of the shipbuilding process; it means that a particular shipbuilding company has been designated by the purchaser to build the ship.
3. *Ship start.* Approximately seven months from the awarding of the contract, metal is cut. Prior to this, however, months have been spent in ship design by naval architects, marine engineers, drafters, and others. Additionally, plans have been made for expansion of the labor force, purchase of equipment, and sequence of construction.
4. *Slab start.* The first metal cutting is fitted together into modules. Simultaneously, piping, ducts, and wiring are started.
5. *Keel laying.* The keel formally serves as the backbone of the ship, and a ceremony accompanies its laying. This usually takes place seven months from slab start. Following this, the hull construction continues as each section is added to the vessel.
6. *The launch.* The first test of a ship comes at launching, when it is about 80 percent completed. With great ceremony, a bottle of champagne is broken on the ship's prow and the ship slides into the water or is floated in dry dock.
7. *Dry-docking.* Following the launch, the ship moves to its outfitting dock for installation of its remaining systems and equipment. While in dry dock, the hull is inspected for seepage or leaks. Dry-docking consists of placing the vessel in a floating structure that permits access to the underwater portion for inspection, cleaning, and repairs.
8. *Light off.* This step occurs when the boiler, engines, and generators are ready to be run and tested.

9. *Builder's trials.* After the launching and outfitting, the ship undergoes a series of tests at sea. A crew—along with personnel from the shipbuilder's office—conducts this shakedown cruise.
10. *Acceptance trials.* Experts representing the supplier and the purchaser inspect, examine, and test the ship and its systems. The inspectors put together a list of the ship's deficiencies. During the following weeks corrections are made.
11. *Delivery.* Once the purchasing company is satisfied, delivery is made. The purchaser may make certain enhancements.

The total amount of time involved in the shipbuilding process may be less than a year for a small vessel. For the largest supertanker, construction may take two and a half years, or more.

The relationship between the builder and the purchaser continues until the expiration of all items under warranty. Deficiencies related to faulty yard work will be corrected.

Shipbuilding and ship repairing employ many workers, including:

boilermakers	naval architects
burners	painters
carpenters	pattern makers
chippers/caulkers	pipe fitters
cleaners	production engineers
clerks	purchasing agents
coppersmiths	riggers
cost accountants	safety inspectors
crane operators	sand blasters
drafters	sheet metal workers
electricians	shipfitters
guards	stenographers
inspectors	storekeepers
machinists	truckers
marine engineers	welders
mold loftsmen	x-ray technicians

The modern American shipbuilding industry constructs ships with a very high level of sophistication. The majority of the industry's employees fall into the category of professional or highly skilled men and women. The following are examples of some of these maritime professionals.

NAVAL ARCHITECTS

Naval architects prepare designs, specifications, and working plans for commercial merchant vessels, naval ships, docks, oil-drilling rigs, and marine facilities often in cooperation with civil engineers. They make cost estimates and supervise the construction of the ship. They advise on problems occurring during building or repair. Naval architects should emphasize the sciences, mathematics, computer programming, and mechanical drawing. Good communication (written and oral) and interpersonal skills are essential to this work because of the diversity of planning and teamwork required. Emphasis of college courses should be in marine engineering or naval architecture, possibly with graduate work in similar areas or in business. Consult your school counselor and college guides for a list of colleges that offer course work in this area.

If you wish to major in naval architecture, the following nine schools offer specific course work in this area:

Department of Naval Architecture
and Marine Engineering
University of New Orleans
New Orleans, LA 70148

Department of Ocean Engineering
Florida Institute of Technology
Melbourne, FL 32901

Massachusetts Institute of
Technology
Cambridge, MA 02139

State University of New York
Maritime College
Fort Schuyler, NY 11542

United States Coast Guard
Academy
New London, CT 06320

United States Naval Academy
Annapolis, MD 21402

University of California
Berkeley, CA 94720

The University of Michigan   Webb Institute of Naval
Ann Arbor, MI 48104   Architecture
                                        Glen Cove, NY 11542

One of the world's most unique colleges, Webb Institute, located in Glen Cove, Long Island, New York, offers a Bachelor of Science in Naval Architecture. All undergraduate students attend the college on scholarship and placement following graduation is 100 percent! It also offers a degree in marine engineering.

Typical of the specialized and sophisticated courses taken by students of Webb Institute's naval architecture programs are the following:

Naval Architecture III—Ship Resistance and Propulsion. Wake fraction, thrust deduction, and propulsive coefficient are presented, and the screw propeller is considered as a propulsion device.

Naval Architecture VII—Ship Design I. The design of a ship to meet a specific set of owner's requirements is undertaken. Principal dimensions, form characteristics, powering, stability, freeboard, and capacity are all considered.

The web site for Webb Institute is: www.webb-institute.edu/htm

Working conditions, salary, and fringe benefits are attractive. Those interested should be able to meet deadlines, concentrate for long periods of time, lead others, make decisions, and exercise patience.

The employment outlook is generally good; new developments in ship construction and maintenance contribute to opening new doors. Private shipbuilders, the federal government (Navy), and naval architecture and marine engineering consulting firms are the main employers.

For additional information, write the following:

| The Society of Naval Architects and Marine Engineers One World Trade Center, Suite 1369 New York, NY 10048 | American Society of Naval Engineers 1452 Duke Street Alexandria, VA 22314 |

Architectural programs do exist for boats. The yacht and small craft curriculum at Chapman School of Seamanship (Stuart, Florida) has courses such as "Hull Design and Construction," "Propulsion and Auxiliary Systems," "Support Systems," and others.

A minor renaissance has occurred in the area of yacht and boatbuilding; North America specializes in these areas. Consequently, we see the expansion of companies such as Cape Cod Shipbuilding—www.four.net/~ccsb/ccsbhp.html. For general information on the boating industry—news, boat shows, and related links—consult www.boatbiz.com.

MARINE ENGINEERS

Closely related to the work of naval architects is that of *marine engineers;* the two professionals often work together as a design team. However, the field of marine engineering is broader. It may include work on ships, oil rigs, and dock construction as well as responsibility for maintaining all the machinery on a ship as a licensed ship's officer. In the latter capacity, a marine engineer starts out as a third assistant engineer and works up to the highest grade, chief engineer, in seven to ten years. Some shipyard-related tasks of marine engineers involve supervision of ship repairs, evaluation during shakedown cruises, and design of boilers as well as propulsion and ventilating systems.

Salaries and working conditions of marine engineers are generally comparable to those of naval architects, as are the necessary personality and academic abilities. The employment outlook for marine engineers is variable. The colleges listed in this chapter for naval architects also offer major programs in marine engineering with the exception of the University of California at Berkeley. The state and national maritime academies also offer programs.

The National Marine Engineers Beneficial Association (AFL-CIO), 30 Montgomery Street, Jersey City, New Jersey 07302, represents approximately ten thousand of its members serving aboard ships. This union and participating companies administer the Calhoon MEBA (Marine Engineering Beneficial Association) Engineering School. In its three-year upgrading program, the school aims to train marine engineers who will maintain and operate all machinery and all systems aboard ship, including sophisticated electronic equipment. Additional information may be obtained from the school:

The Calhoon MEBA
    Engineering School
    Route 5
    Easton, MD 21601

SKILLED CRAFTS

This section summarizes many of the crafts that compose the skilled workforce in the shipyard.

One of the most important operations in the yard is *welding,* the technique that joins the metal pieces of the ship together. *Metalworkers* of all kinds heat, cut, bend, and shape their materials into parts that will be welded together. As the ship takes form, *carpenters* prepare a building basin or building ways for the laying of the keel and the subsequent assembly.

Crane operators use huge cranes to lift preassembled parts into place. Prefabrication takes place in the shops in the yard, thus reducing cost. Pennsylvania Shipbuilding (Chester, Pennsylvania) has a unique floating crane, with eight hundred tons capacity. It's a virtual twenty-three-story skyhook.

Shipfitters see that each piece is correctly located and securely fastened by *riveters, welders,* and other workers. Edges are trimmed by *chippers* by air arc process.

A new staff takes over following launch, when the ship moves to the outfitting pier. At this point, *engineers* and *machinists* prepare, install, and test propulsion equipment. *Electricians, carpenters, boilermakers, sheet-metal workers, painters, plumbers,* and others construct internal systems and facilities.

If you want to get skilled work in the yard, you should pursue high school courses related to your special interest (electricity, machinery, mechanical drawing, computer-aided drafting and design, carpentry). Employers prefer men and women who have graduated from high school and have had some work experience. Once employed, the new worker begins as a helper (for example, a plumber's helper aids and learns from a master plumber), then proceeds to a formal apprenticeship program for three or more years. Following the apprenticeship, the employee qualifies as a journeyman (not included in all crafts). The final level is *master,* meaning a fully skilled worker.

Shipyard employees, as skilled workers, earn a good salary and enjoy good benefits. Some specialists with skills receive premium salary. The employment outlook remains fair, since certain segments of the shipbuilding industry are undergoing a mild renaissance. A variety of unions dominate the industry and operate apprenticeship programs in cooperation with shipbuilding companies. For additional information contact the following firms (because of the state of shipbuilding, some "yards" have merged and others have closed and these trends will continue):

Alabama Drydock and
   Shipbuilding Company
P.O. Box 1501
Mobile, AL 36601

The American Shipbuilding
   Company
911 Bond Court Building
Cleveland, OH 44114

Avondale Shipyards Inc.
P.O. Box 50280
New Orleans, LA 70150

Bath Iron Works Corp.
700 Washington Street
Bath, ME 04530

Bethlehem Steel Corp.
Shipbuilding Division
Sparrows Point Yard
Sparrows Point, MD 21219

General Dynamics Corporation
Pierre Laclede Center
St. Louis, MO 63105

Jacksonville Shipyards, Inc.
750 East Bay Street
P.O. Box 2347
Jacksonville, FL 32203

Litton Industries, Inc.
Ingalls Shipbuilding Division
P.O. Box 537
Pascagoula, MS 39567

National Steel and Shipbuilding
Company
P.O. Box 80278
San Diego, CA 02138

Norfolk Shipbuilding and Drydock
Company
P.O. Box 2100
Norfolk, VA 23501

Pennsylvania Shipbuilding
Company
P.O. Box 498
Chester, PA 19013

Rohn Marine, Inc.
P.O. Box 2300
Chula Vista, CA 92012

Tenneco/Newport News
Shipbuilding
14101 Washington Avenue
Newport News, VA 23607

Todd Shipyards Corporation
Jersey City, NJ 07303

Additionally, the United States Navy operates several shipyards that employ civilians, such as the one at Pascagoula, Mississippi, web site: www.sspasc.navy.mil/

The U.S. Maritime Administration (MARAD) is charged with the mission "to promote the development and maintenance of an adequate, well-balanced, United States merchant marine, sufficient to carry the Nation's domestic waterborne commerce and a substantial portion of its waterborne foreign commerce, and capable of serving as a naval and military auxiliary in time of war or national emergency." MARAD also "seeks to ensure that the United States enjoys adequate shipbuilding and repair service, efficient ports, effective intermodal water and land transportation systems, and reserve shipping capacity in time of national emergency." See its web site at: http://marad.dot.gov/ and http://marad.dot.gov/linksshipyard.html.

CHAPTER 8

MILITARY CAREERS AFLOAT

UNITED STATES NAVY

Armed with awesome technology, the Navy of the twenty-first century will be smaller in both vessels and personnel. Faced with new demands and many traditional responsibilities, the U.S. Navy continues to offer men and women a rewarding career with wonderful possibilities either for a career of military service or for just one "hitch."

Enjoying its third century of service and commitment to the American people, the United States Navy began its career during the American Revolution. The function of the Navy during wartime is well known; however, Americans give little thought to its peacetime activities. The Navy has successfully responded to both tasks. It has distinguished itself in a series of military encounters with foreign powers, including the two great challenges of this century.

Although considerably smaller than it was with 6,000 ships in its service at the end of World War II in 1945, today's Navy, numbering approximately 340 ships, remains vigilant. More diverse than its earlier counterparts, its fleet includes not only mammoth aircraft carriers but small boats capable of speeds over one hundred miles per hour. Navy personnel number about 380,000, of which 55,000 are officers and 325,000 are enlisted people; some others are in college NROTC programs and at the U.S. Naval Academy. Peacetime naval duties include a great variety of research activities such as shark and dolphin behavior, undersea habitat, weather forecasting, weapons development, interpersonal relationships, ship development, space exploration, water pollution prevention, and laser technology. The Navy remains available for evacuation in case of natural disasters and political crises and for deployment around the world to maintain security. During noncombat periods, the Navy practices a variety of tactics and exercises its weaponry and personnel.

Given its many roles, the Navy is a large and complex organization with tremendously varied career opportunities. Currently, there are more than seventy-five job skills open to men and women. Navy career positions may be divided into enlisted personnel and officers.

ENLISTED PERSONNEL

When new enlistees enter the Navy, they usually receive the position of *seaman recruit*. Advancement is through the ranks in the following order:

seaman recruit
seaman apprentice
seaman
petty officer, third class
petty officer, second class
petty officer, first class
chief petty officer
senior chief petty officer
master chief petty officer

The Navy currently provides eight and a half weeks of recruit training with a full schedule of military and technical subjects. All

inspecting lifesaving equipment. Obviously, an interest in tools and machines is important, as well as manual dexterity and the ability to perform detailed work. Only qualified applicants are accepted into this rank. Construction engineers are helped by *engineering aides.* The latter conduct surveys, prepare blueprints and sketches, and lay out all kinds of construction. Once the actual construction work begins, they aid in maintaining quality control until the project's completion.

The expansion of computer systems to aid in information utilization has not escaped the Navy. *Data processing occupations* fall into two ratings: *data processing technician* and *data systems technician.* The former operate computers, write programs for processing operations, and perform office management functions. The latter adjust and repair digital computers, video processors, tape units, buffers, key sets, digital display equipment, data link terminal sets, and related equipment. As with other careers requiring technical training, assignment may be shore or ship duty.

Health occupations include not only doctors and dentists, but also technicians and assistants. There are many opportunities for careers in this area. For example, hospital corpsmen assist medical professionals in providing health care to service personnel and their families. Specifically, they apply first aid, minister to the sick and injured, take and process x-rays, and perform similar health-related duties. A science background proves very helpful. Similarly, *dental technicians* assist dental officers in treating patients. They arrange appointment schedules, instruct patients in oral hygiene, assist in surgery, and prepare materials for fillings.

Manufacturing occupations employ skilled and semiskilled workers. Skilled labor requires considerable training, while unskilled workers normally learn on the job. Only a small number of sailors are used by the Navy as *molders.* Molders work in foundries in which molten metals are formed into parts, operating and maintaining the ovens. Physical strength and manual dexterity are important qualifications. Future prospects are good. *Steel-*

workers utilize special equipment to construct or demolish steel structures. They erect or dismantle bridges, buildings, piers, and similar facilities. In performing these tasks, special equipment such as the oxyacetylene blowtorch is used. Those interested in this job should be able to handle tools, equipment, and machines. Career opportunities appear to be good.

Among the fastest growing fields, both in the Navy and society in general, are *mechanical* and *repair occupations.* Workers learn their skills on the job or through formal apprenticeship training. Men and women who inspect, adjust, test, repair, and overhaul aircraft engines are rated as *aviation machinist's mates.* They work on both helicopter and jet engines. Excellent opportunities exist in this career area. *Machinist's mates* maintain and repair ship propulsion equipment such as propellers, pumps, oil purifiers, and reduction gears. They also maintain elevators, air-conditioning equipment, and steering mechanisms. Those interested in mechanical and repair occupations should have the ability to work with machinery and equipment and the strength to perform physical tasks.

Scientific and technical occupations ratings account for a very large percentage of careers in the Navy. Some examples are *aviation antisubmarine warfare technician, missile technician, ocean systems technician,* and *torpedoman's mate.*

One of the important peacetime contributions of the Navy lies in weather forecasting. It is the *aerographer's mate* who monitors temperature, wind speed, humidity, and other factors used to predict the weather. Once the data have been collected, they must be plotted and analyzed, often with the help of computers.

Gunner's mates are responsible for the operation and maintenance of guided missile systems, rocket launchers, gun mounts, small arms, and similar armaments. *Operations specialists* operate radar, navigation, and communication equipment in the combat information center near the ship's bridge. Persons in these positions identify ships and aircraft and determine their distance, speed, and altitude. Scientific and technical occupations are probably among the most glamorous in the Navy—the kind frequently shown in recruiting films. Opportunities remain good to excellent.

A wide range of careers falls under the heading of *service occupations.* Included here are cooks, barbers, police officers, and building custodians. A *mess management specialist* operates kitchen and dining facilities, ordering foodstuffs, preparing menus, and serving meals. Some supervisory people manage the officers' and enlisted personnel's quarters. The management of barber, tailor, and uniform shops comes under the function of the *ship's servicemen,* who also administer the recreation clubs, retail and commissary stores, and laundry and dry-cleaning plants. These responsible positions require skills in keeping accurate records and dealing with people.

Workers in *social science occupations* specialize in aiding people with their problems of daily living. For example, *legalmen* and *legalwomen* provide assistance in matters such as voter registration, indebtedness, wills, taxes, customs, legal forms, and claims filing. Likewise, men and women *career counselors* interview personnel, administer tests, and make recommendations for assignments. They suggest career opportunities and avenues of advancement. In all people-oriented ratings, facility with language and empathy with people are important assets.

Transportation field occupations encompass a large number of naval ratings. Traffic control at airfields and on aircraft carriers is directed by *air controlmen* and *air controlwomen.* These specialists coordinate the safe movement of aircraft with radar information and voice radio. They must remain alert to visibility, weather, and lighting systems. Good eyesight and speaking voice are critical tools to this occupation.

One of the oldest jobs in the Navy is the *boatswain's mate,* who keeps the outside surfaces of ships in good condition, maintains machinery and equipment on ship's decks, handles cargo, and operates barges, tugs, and small boats. This crew member also supervises deck crew in cleaning and painting and teaches seamanship. In addition to these functions, part of the watch standing duties includes the actual steering of the ship—in fact they drive the ship!

For the enlisted person, the Navy offers many benefits, not the least of which is training. However, it behooves the prospective recruit to gain all the education possible prior to enlisting. After service in the Navy, whether for one hitch or a career, many opportunities become available in civilian life. Most private companies are anxious to hire persons with naval training and experience.

THE NAVAL OFFICER

The naval officer serves in a position of leadership in the Navy. Consequently, a fine liberal studies, professional, and technical education is particularly important. Most officers obtain their training through the United States Naval Academy, NROTC (Naval Reserve Officers Training Corps) programs, Aviation Officer Candidate Schools (AOCS), or Officers Candidate School (OCS).

The Naval Academy

The United States Naval Academy takes as its mission "to prepare midshipmen morally, mentally, and physically to be professional officers in the naval service." The academy is located in Annapolis, Maryland, a beautiful nautical setting near the Chesapeake Bay. The curriculum consists of a typical college program coupled with professional military and leadership training. The variety of majors has expanded greatly in recent years. Seven designated bachelor of science degrees are offered in aerospace engineering, electrical engineering, mechanical engineering, systems engineering, marine engineering, naval engineering, and ocean engineering. Students at the academy may also elect undesignated Bachelor of Science degrees with majors in the following fields:

aeronautical engineering	marine engineering
astronautical engineering	mathematics
chemistry	mathematics honors
computer science	mechanical engineering
economics	naval architecture
economics honors	ocean engineering
electrical engineering	oceanography
English	oceanography honors
English honors	physics
general engineering	political science
general science	political science honors
history	quantitative economics
history honors	systems engineering

The following course descriptions provide a sample of the technical matter studied by midshipmen and midshipwomen:

Principles of Aeronautics and Flight. An introductory course in the theory and practice of atmospheric flight, propulsion, rocket performance, space flight, satellite systems and uses, and other topics in the field of astronautical engineering. It is designed to prepare the beginning aerospace engineering student for future studies and to develop good engineering practices.

Naval Engineering. An introduction to ship systems, including basic methods of ship procurement, construction, and power plant selection. Principles of ship stability and operability as related to preventive and corrective damage control.

Ship Structures. A course in structural theory and practice. Topics include longitudinal and transverse strength of the hull girder, bending moments in a seaway, plate theory, development of ship structural design, submarine pressure hull design, and shipbuilding materials.

Seakeeping and Maneuvering. Topics include ship steering, maneuvering, motion, and seakeeping. The basic equations of motion for a maneuvering ship and for ship motions in a seaway are developed, and various methods of solution are discussed. The course also covers the experimental aspects of sea keeping and maneuvering.

Fundamentals of Naval Science. Introduction to the basic concepts of seamanship, ship handling, coastal piloting, and engineering. Includes at-sea labs on 108' Yard Patrol Craft (YPs).

Polar Oceanography. A descriptive course that covers the history of polar exploration as well as the physical oceanography and meteorology of the polar regions with particular emphasis on the role of sea ice in global warming studies. Current DoD polar programs will be reviewed including the following: Deep Freeze, International Ice Patrol, and Navy operations such as ICEX and TEAMWORK.

Following the four years of study, the graduate must serve a five-year commitment as an active duty officer. Additional information may be obtained by contacting the academy:

Director of Candidate Guidance
 United States Naval Academy
 117 Decatur Road
 Annapolis, MD 21402-5018
 (800) 638-9156

Additionally, the academy maintains an excellent web site at: http://www.nadn.navy.mil/

NROTC

Many colleges and universities prepare naval officers through NROTC (Naval Reserve Officers Training Corps) programs. These midshipmen and midshipwomen lead essentially the same campus life as civilian undergraduates. In addition to their normal studies, they must satisfactorily complete specified naval science courses and participate in scheduled drills and summer training periods. Some students receive scholarships and a stipend; in addition, nonscholarship students receive an allowance in their junior and senior years. Upon satisfactory completion of all requirements, candidates receive commissions and must serve four years as officers. For additional information contact the following schools:

Auburn University
Arizona University
Boston University
California, University of (Berkeley)
California, University of (Los Angeles)
The Citadel
Colorado University
Cornell University
Duke University
Florida, University of
Florida A & M University
George Washington University
Georgia Institute of Technology
Hampton University
Holy Cross, College of
Idaho, University of
Illinois, University of
Illinois Institute of Technology
Iowa State University of Science and Technology
Jacksonville University
Kansas, University of
Maine Maritime Academy
Marquette University
Massachusetts Institute of Technology
Memphis, University of
Miami University (Ohio)

Michigan, University of
Minnesota, University of
Mississippi, University of
Missouri, University of
Morehouse College
Nebraska, University of
New Mexico, University of
Norfolk State University
North Carolina, University of
North Carolina State University
Northwestern University
Norwich University
Notre Dame, University of
Ohio State University
Oklahoma, University of
Old Dominion University
Oregon State University
Pennsylvania, University of
Pennsylvania State University
Prairie View A & M University
Purdue University
Rensselear Polytechnic Institute
Rice University
Rochester University
San Diego, University of
Savannah State College
South Carolina, University of
Southern California, University of
Southern University A & M College
State University of New York Maritime College
Texas A & M
Texas, University of
Tulane University
Utah, University of

Vanderbilt University
Villanova University
Virginia, University of
Virginia Military Institute
Virginia Polytechnic Institute and State University
Washington, University of
Wisconsin, University of

Many of the above colleges have "cross town" relationships permitting students at other colleges to become part of the program at a NROTC school. For more information contact one of the above listed institutions of higher education.

OCS

Qualified graduates of regionally accredited colleges receive sixteen weeks of training at Officers Candidate School. Located in Newport, Rhode Island, and Pensacola, Florida, these schools provide basic naval science and indoctrination; successful graduates are commissioned as ensigns. More information may be obtained by calling (800) USA-NAVY or on the web at http://www.navyjobs.com.

Naval officers have interesting and challenging careers, both on shore and at sea. Some engage in high-level security research at the Pentagon; others instruct future officers; still others command nuclear submarines. Officers advance through the following ranks:

ensign
lieutenant (junior grade)
lieutenant
lieutenant commander
commander
captain
rear admiral

vice admiral
admiral

The organization of personnel aboard ship closely resembles that of a war organization. The basic shipboard departments are navigation, operations, weapons or deck, engineering, and supply. The *commanding officer* has absolute authority and responsibility for the ship, subject to the limitations of law and naval regulations. The commanding officer exerts every effort to maintain the ship and its personnel in a state of maximum readiness. During military action the commanding officer must engage the enemy and fight to the utmost of his or her ability. If the commanding officer is absent or unable to fulfill these duties, the next eligible officer in grade assumes command.

The *executive officer* functions as an aide or assistant to the commanding officer. As second in command, this person contributes to maintaining the ship's routines efficiently. Specifically, the executive officer assigns personnel and maintains their records, supervises loading and berthing plans, and oversees morale, welfare, and discipline. The executive officer has several assistants who help plan and perform the work, including the *ship's secretary,* who is responsible for the administration of the ship's correspondence and personnel records, and the *legal officer,* who functions as an advisor and staff assistant to the commanding and executive officers.

Each of the ship's departments is headed by an officer. For example, the *engineering officer* is charged with the operation, care, and maintenance of the propulsion and auxiliary machinery, damage control, electrical devices, repairs to the hull, and similar jobs. Among the several assistants is the *engineering officer of the watch,* who reports any defects of machinery that may affect the proper operation of the ship. Several other officers serve aboard ship in other functions. More specific information may be obtained through the official web site of the U.S. Navy: http://www.navy.mil/.

Two additional organizations can provide quality information on the Navy—the U.S. Naval Institute and the Naval Historical Foundation. The Naval Institute publishes numerous books and conducts programs relative to their mission, "the advancement of professional, literary, and scientific knowledge in the maritime services, and the advancement of knowledge of sea power." The Naval Institute may be accessed at: http://www.usni.org/. The Naval Historical Foundation focuses upon the history of the Navy with attention to many of its leaders, battles, and related events. Its web site is: http://www.history.navy.mil/.

THE COAST GUARD

The United States Coast Guard, often called the Navy of Mercy, has had a long and distinguished career. Beginning in the early years of the country as the Revenue Marine, it sought to dissuade the flourishing smuggling trade. Today's Coast Guard came into existence prior to World War I with the merger of the Revenue Marine, the Life Saving Service, and the Lighthouse Service. These latter two functions, generally believed to be the most colorful and glamorous, still form an important part of the Coast Guard's work. Contemporary lifesaving methods utilize highly skilled personnel and the modern technology of speedboats and helicopters. The Coast Guard also takes care of more than forty-four thousand buoys and channel markers, detects and monitors icebergs, collects oceanographic data, enforces drug interdiction, fights to control maritime pollution, maintains port safety and security, clears ice-blocked shipping lanes, administers safe pleasure-boating programs, enforces the two-hundred-mile Offshore Fish Conservation Act, and maintains military readiness. During an average day, the U.S. Coast Guard will:

- board 90 large vessels for port safety checks
- process 120 seamen's documents
- seize 209 pounds of marijuana and 170 pounds of cocaine worth $9.2 million
- conduct 120 law enforcement boardings
- investigate 17 marine accidents
- inspect 64 commercial vessels
- save 14 lives
- assist 328 people in distress
- save $2,490,000 in property
- service 150 aids to navigation
- interdict 176 illegal migrants

The Coast Guard numbers some thirty-five thousand active duty men and women, eight thousand reservists, and thirty-five thousand auxiliaries. The general ratings categories are as follows:

- deck and ordnance group
- hull and engineering group
- aviation group
- administration and scientific group

Enlisted personnel attend boot camp in Cape May, New Jersey; following this experience, they enter training programs for a number of specialties. One of the most responsible positions within the deck and ordnance group is that of *quartermaster.* This person reads the charts, signals, and other measurements that affect the vessel's navigation. The safety of the vessel depends upon the quartermaster's skill. In addition to above-average leadership ability, good vision and hearing and strong communication skills are important. *Sonar technicians* use sonar equipment to locate submerged objects, reefs, and submarines. They also maintain this equipment.

Several technical specialties fall into the hull and engineering category. For example, one of the fastest-growing jobs in the Coast Guard is that of the *electronics technician,* who handles the continuous checking and repair of sophisticated electronic equipment such as radar, sonar, and radio receivers and transmitters. *Machinery technicians* work on internal combustion engines, refrigeration, air-conditioning, propulsion machinery, and other equipment.

Within the aviation group one finds all the ratings relative to aircraft and their functions. The *aviation electrician's mate* works with basic wiring, power circuits and motors, and the maze of electrical systems aboard aircraft and shore installations. The maintenance of parachutes, small arms, munitions, search and rescue equipment, and similar items falls to the *aviation survivalman.*

The administration and science group includes *yeomen, public affairs specialists, storekeepers,* and *radiomen.* Others in this category are the *port securityman,* who maintains and controls the safe transfer of dangerous cargoes, and the *marine science technician,* who observes, collects, analyzes, and disseminates meteorological and oceanographic data.

The Coast Guard maintains an academy (New London, Connecticut) and officer candidate school (Yorktown, Virginia) similar to those of the Navy. Courses include liberal arts, engineering, and maritime specialties, with special emphasis on the history and mission of the Coast Guard. The beautiful sailing ship, the *Eagle,* is used as a training vessel. Upon graduation, the new Coast Guard officer may receive one of several assignments, from the poles to the tropics. Some typical assignments are as follows:

aviation	merchant marine safety
boating safety	military readiness
civil engineering	naval engineering
data processing	personnel administration
drug interdiction	port safety
environmental protection	public affairs
intelligence	reserve administration
legal assistance	search and rescue

For further information on Coast Guard officer training, write:

Director of Admissions
  United States Coast Guard Academy
  15 Michigan Avenue
  New London, CT 06320-9807

For the interested individual, the Coast Guard can serve as the avenue to an exciting and rewarding career with good financial security. As with the Navy, advancement in rank brings commensurate salary increases. (For specific officer ranks, see the earlier section in this chapter on the Navy.) Pension plans are attractive and former Coast Guard men and women have many opportunities in the business world because of their skills and experience.

The Coast Guard also has a program that high school students can enter during the summer of their junior year. For additional information on this program, consult the yellow pages for your local recruiting office or contact the following:

U.S. Coast Guard Recruiting Center
  4200 Wilson Boulevard, Suite 450
  Arlington, VA 22202
  (800) 424-8883
  (800) GET-USCG (to talk to your local recruiter)
  http://www.uscg.mil/hq/recruit/recruit.htm

Those interested in opportunities relative to the Canadian Coast Guard should contact:

Canadian Coast Guard
  Department of Fisheries and Oceans
  200 Kent Street
  Ottawa, CA K1A 0E6
  (613) 993-0215
  http://www.ccg-gcc.gc.ca/

In addition the Canadian Coast Guard College located in Sydney, Nova Scotia maintains an excellent web site for the college and for related topics at: http://www.cgc.ns.ca/.

CHAPTER 9

MISCELLANEOUS MARINE AND MARITIME CAREERS

To provide a sense of the breadth and variety of maritime occupations across North America, this chapter describes a variety of marine and maritime jobs. These jobs offer interesting and independent work in addition to adequate remuneration. It is hoped that, from the material in this chapter, you will come to understand that a worthwhile marine or maritime career can be had anywhere, at many different levels, under all kinds of conditions—if you have the skill and creativity to find or create it.

MARINE TECHNICIANS

One of the career categories receiving renewed attention in the contemporary job market is that of the *marine technician.* To quote from a recent report prepared by the staff of Monterey Peninsula College, *A National Forum on Critical Issues in Marine Advanced Technology Education:*

> Marine technicians are individuals who apply science and engineering skills to the marine environment, including the open ocean, coastal regions, estuaries, rivers, swamps, and lakes. They may work on board ships or other vessels, directly underwater (e.g., diving, in submarines), remotely underwater (e.g., ROVs), in a marine laboratory, or in any

number of other marine and coastal settings. Their work often demands the ability to function safely and efficiently in the marine environment.

These individuals work in a wide range of jobs including,

aquaculture technician
boatbuilding assistant
commercial deep-sea diving specialist
hyperbaric technician
marine engine mechanic
oceanography technician
submersible operator

Several of these occupations have received coverage in other chapters in this book; one career deserves special attention here—that of *marine engine mechanic.*

While still in high school, George Nylund began working at the Tuckerton Marina in New Jersey. He worked for two summers, as well as after school and on weekends during the school year. In that time he advanced from small clean-up jobs around the marina to tune-ups on the motorboats. Upon high school graduation, George decided to forego college and open Tuckerton Marine Service. His small company would perform maintenance and repair tasks. Just a couple of years old, his company has a bright future. Most investments of this nature take at least three years to show a profit. Over the late spring and the summer, George works long hours, as do his two part-time assistants. However, during the off-season, he makes his own hours, except for attending the training programs conducted by the various marine engine organizations.

Owning your own company, particularly one in which you have the opportunity to work close to water with a variety of boats, makes for an enjoyable life. However, such an undertaking requires good business acumen (accounting, taxes, government regulations), interpersonal skills, and leadership ability. The latter involves supervising people working for you in a considerate yet structured manner. Leadership also involves decision making in terms of purchasing equipment and materials, expanding the company, advertising, and similar endeavors.

Pleasure-boating industry sales expand yearly. Annually, millions of Americans use more than ten million outboards. At present, more than six thousand marinas, boatyards, and yacht clubs serve American boaters. Some eighteen hundred boatbuilding firms exist in the United States. Some employ hundreds in their many plants.

This is a major industry requiring many mechanics to service these boats. For specific addresses of boatbuilding companies, please consult the following:

Boating Industry Magazine
Buyers Guide Issue
850 Third Avenue
New York, NY 10022

The industry has expanded so greatly that several colleges in coastal areas and near large bodies of water have begun to offer programs for marine engine mechanics. Typical of these is Florida Keys Community College, which offers a marine diesel technology program. Diesel power is prominent in marine propulsion. Commercial shrimp and fishing fleets, charter boats, ferry boats, yachts, and tugboats typically utilize diesel power. Students can earn a certificate in marine gasoline engine technology or marine diesel technology. The latter is a one-year, thirty-one-semester-hour program that leads to a certificate in marine diesel technology. The school also offers a two-year program in marine propulsion technology leading to an associate degree.

Graduates of both programs take jobs as charter-boat captains, diesel mechanics, master engineers for commercial fishing vessels, and service managers for marine equipment suppliers.

Another Florida school offering several programs for marine technician is the Chapman School of Seamanship in Stuart. One of the programs relative to this topic is Boat and Engine Maintenance, which requires eighty-four hours of instruction. The curriculum focuses on three short courses: Boat Maintenance, Marine Mechanical Systems, and Vessel Systems.

According to a recent *Sea Technology Buyers Guide,* one of the major problems facing Gulf of Mexico workboat owners is the lack of qualified personnel; so the future for some job specialties looks good.

Community colleges and private proprietary schools (for profit) have taken the lead in the marine technology area. Most feature a combination of classroom and practical experience. Some of these schools follow:

Alexandria Technical College, Minnesota
Bellingham Technical College, Washington
Cape Fear Community College, North Carolina
Chapman School of Seamanship, Florida
Clatsop Community College, Oregon
Clover Park Technical College, Washington
Coastal School of Deep-Sea Diving, Inc., California
College of Marin, California
College of Oceaneering, California
Diver's Academy of the Eastern Seaboard, New Jersey
Divers Institute of Technology, Washington
Florida Keys Community College, Florida
Gray's Harbor College, Washington
Great Lakes Maritime Academy, Michigan
Gulf Coast Community College, Florida
Indian River Community College, Florida
Maine Maritime Academy, Maine
Maritime Institute of Technology, Maryland
North Georgia Technical Institute, Georgia
Peninsula College, Washington
Prince William Sound Community College, Arkansas
Saddleback College, California
Santa Barbara City College, California

Seattle Central Community College, Washington
Shoreline Community College, Washington
Southern Maine Technical College, Maine
State University of New York, Morrisville, New York
Suffolk County Community College, New York
University of the District of Columbia, District of Columbia
Washington County Technical College, Maine

Monterey Peninsula College (MPC) in Monterey, California, has spearheaded a program to improve the education and training of personnel in this field through working with institutions and organizations. The Marine Advanced Technology Education (MATE) Center at MPC (founded in 1997) hopes to organize a national network to coordinate the education of marine technicians. They plan to disseminate information to benefit students and faculty. For additional information contact:

MATE Center
  Monterey Peninsula College
  980 Fremont Street
  Monterey, CA 93940
  http://www.marinetech.org

If you like being around the water and have a corresponding interest in working with gas engines, visit a boat basin, yacht club, or marina. After walking around the docks and talking with some people, ask for work. Whether or not you are able to gain experience, you may wish to write to the companies that build pleasure-boat engines. In your letter, request information on them, their products, and potential employment.

NAUTICAL ARCHAEOLOGISTS

One of the most exotic of the occupations in the world of water involves treasure hunting under the sea. This recent science has come to the attention of Americans because of the interest stirred by the discovery of a number of sunken ships and treasures. Origins of this field of study date to the year 1900, when a party of Greek sponge divers accidentally found an ancient shipwreck and salvaged priceless statues from it. For many years, marine archaeology remained a crude science; often wrecks were pirated for their booty with little attention to historical artifacts.

The development of SCUBA equipment, undersea vehicles, and systematic procedures for rescuing artifacts from the sea and reconstructing them on shore has contributed to advancing the academic discipline of nautical archaeology. One author and undersea archaeologist describes an experience:

> Before ascending, I took a long look around at what undoubtedly was the ruin of a ship. What ship was she? I wondered. Out of what port? Where was she bound, carrying what cargo, and under whose hand? What adventures had she survived before the final one that brought her to the bottom?

The first step in the challenge involves the awesome problem of locating the ship. Once the assembled team has a general idea where a ship may be found, it then proceeds to find the vessel. Many times this becomes a difficult task resulting in failure. If discovered, following the initial examination, a grid of pipes is laid out in the area of the ship. As each piece of wreckage is brought to the surface, it is labeled to facilitate later reconstruction. This field involves serious and painstaking procedures. History thus salvaged from the sea includes items such as cannons, cutlasses, religious objects, water jugs, eating utensils, and even exotic items like lime juicers and dice. Archaeologists use these artifacts and information about the period to reconstruct an idea of the ship, its mission, and the people who formed the society of which the sailors were members. Consequently, marine archaeologists contribute to a greater understanding of the past.

Carmen Marquez-Marin operates her own business, Archeo Marine Research, in San Juan, Puerto Rico. In an interview for *Marine Science Careers,* she told of her love for her occupation. "I like having the opportunity to combine my interest in human behavior and my love of the sea while studying how human beings have interacted with the marine environment at different times in history."

The outlook for marine archaeology appears to be good because of a growing general interest in the field and the continued development of undersea technology. However, few opportunities exist; most undersea archaeologists are employed only part-time. Many work in museums and serve on university faculties. Those interested in entering the profession would be well advised to obtain an advanced degree. Some supportive positions are filled by diving specialists, who are hired on a part-time basis.

The following college offers a unique master's degree in Maritime History and Nautical Archaeology.

Admiral Ernest M. Eller House
    East Carolina University
    Greenville, NC 27858-4353

The following universities offer programs that award M.A. and Ph.D. degrees:

Nautical Archaeology Program
    Department of Anthropology
    Texas A & M University
    College Station, TX 77843-4352
    http://nautarch.tamu.edu/

Underwater Archaeology Program
    Florida State University
    Tallahassee, FL 32306
    http://www.adp.fsu.edu/uwarch.html

The above programs have made significant contributions to our knowledge of previous societies and cultures and their activities.

Their faculty, students, and alumni continue to investigate the deep as they shape the future of nautical archaeology.

Other organizations you may wish to contact are:

The Institute of Nautical Archaeology
P.O. Drawer HG
College Station, TX 77841-1537
http://www.nautarch.tamu.edu/ina/history.htm

The International Institute for Maritime Archaeology
The Bermuda Maritime Museum
P.O. Box MA 273
Somerset, MA BX Bermuda

SHIP RESTORERS

Somewhat related to the career field described above is that of ship restoration, which often involves wooden ships of an earlier era. For example, during America's Bicentennial Celebration Year, the voyage of the Tall Ships from Bermuda to New York City, an accident occurred in which the *Gazela Primiero* (aka *Gazela*) was rammed by another ship. Fortunately, the damaged ship was able to be repaired by a ship's carpenter at Mystic Seaport in Connecticut. This incident focused national attention on this career. The carpenter was one of a few people in this country knowledgeable in the skills used decades ago on sailing ships. Several museums are in the process of restoring vessels to their original condition and need persons with sailing ships' crafts. So important are these persons that museums jealously guard those in their employ. Naturally, these restoration experts receive attractive salaries.

The field of wooden boatbuilding is more active than most would expect. Wooden minesweepers are being built in Wisconsin and wooden shrimpers and fishboats in Florida and Washington state. Along the eastern seaboard and in the northwest, wooden sail and power pleasure boats are under construction. Larger fiberglass pleasure boats have increasing amounts of wood on deck and in the interior. Numbers of older wooden pleasure boats need repair and restoration. Although not a large industry, it requires skilled craftspeople familiar with traditional and modern techniques of building and repair.

Bath, Maine, a cradle of boat and shipbuilding activity for more than three hundred years, boasts a wonderful program in ship preservation skills. This branch of the Maine Maritime Museum trains apprentices who spend eighteen to twenty-four months at the site, as well as many educators who spend shorter periods of time onboard. The program ties in nicely with the national trend toward using one's hands and has had an excellent response. The museum's apprentices build wooden vessels as they were built in the last century, utilizing time-tested skills and procedures. For example, oils, pine tar, and turpentine are used as preservatives. Persons interested in this career should contact the following:

Director of Apprenticeship
  Maine Maritime Museum
  Bath, ME 04530

For general information on maritime restoration, write the following:

National Trust for Historic Preservation
  740-48 Jackson Place NW
  Washington, DC 20006

Tri Coastal Marine
  P.O. Box 1049
  Galveston, TX 77555

Training for these careers is available at a number of schools. The Norfolk School of Boatbuilding, founded in 1977, offers courses titled, Introduction to Woodworking Tools and Machinery, Introduction to Traditional Boatbuilding, Understanding Small Craft Design, and others. Additional information may be obtained on training from the following:

The Apprentice Shop
Box 539
Rockport, MA 04856

The Arques School of Traditional
Boatbuilding
P.O. Box 2010
Sausalito, CA 94966
http://www.arqueschl.org/

The Bear Mountain Boat Shop
P.O. Box 191
Peterborough, Ontario
CANADA F9J 6S8
http://www.bearmountainboats.
com

Cape Fear Institute
Wilmington, NC 28401

Landing School of Boatbuilding
and Design
Box 1490

Kennebunkport, ME 04046
http://www.by-the-sea.com/
thelandingschool/

Norfolk School of Boatbuilding
Box 371
Norfolk, VA 23501

Northwest School of Wooden
Boatbuilding
251 Otto Street
Port Townsend, WA 98368
http://www.olympus.net/
woodboat/

Philadelphia Wooden Boat Factory
1107 Wharton Street
Philadelphia, PA 19147
http://www.libertynet.org/pwbf/

Wooden Boat School
Box 78
Brooklin, ME 04616

MARITIME MUSEUM CAREERS

Maritime museums vary greatly; some feature one ship, such as the museum ship *Norgoma* at the St. Mary's River Marine Center in Sault Ste. Marie, Ontario, Canada, and the U.S. Naval Museum/ Submarine USS *Ling* in Hackensack, New Jersey. Many larger museums also exist that feature several vessels, exhibit areas, and a significant staff. Patriot's Point Naval and Maritime Museum (Mt. Pleasant, South Carolina), Independence Seaport Museum (Philadelphia, Pennsylvania), and San Francisco Maritime Historical Park (San Francisco, California) constitute examples of excellent large facilities. North America has many, many examples of museums of excellent quality large, medium, and small (see Appendix B). Positions vary accordingly.

A *curator* serves as the professional caretaker of the museum's collections. In a large museum several curators would be employed, with a *director* serving as the administrator. In a small museum the curator usually also serves as the manager. In recent years, many museums have added an *education director.*

Museums must carefully display their collection to interest and educate the public. Curators and their staffs plan the exhibits of artifacts: ships' anchors, pieces of eight, tools, weapons, photographs, maps, charts, and similar items. Often an organization or individual will lend items for a specific display.

Ship and artifact restoration is an important craft related to museum work. The Naval Academy Museum collection emphasizes the history of the Navy (uniforms, ship models, and flags), and the Patriot's Point Naval and Maritime Museum in Mt. Pleasant, South Carolina, has as its focal point the USS *Yorktown* and the N.S. *Savannah.* This museum plans dramatic expansion of its facilities. The *Alexander Henry,* a three-thousand-ton icebreaker, forms the focus of the museum of the Great Lakes at Kingston in Ontario, Canada.

Maritime museum careers begin with a strong interest in the sea. Some employees are history buffs; others are professionally trained people with master's degrees or Ph.D.s in history; a few have had course work in museum studies.

Some museums, such as the magnificent Mystic Seaport Museum in Connecticut, offer internships in museum studies. Normally these are master's level programs associated with colleges. The work generally takes place indoors and offers opportunities to work with historical artifacts as well as with the public.

Many museums have youth programs. For example, the *Gazela* of Philadelphia has a sea cadet organization. Mystic offers opportunities for sleep overs for school students as does Patriot's Point for Girl and Boy Scouts. They and many others offer summer programs. Activities such as these offer an excellent opportunity to test the waters of this profession.

In Appendix B you will find a list of the locations of museums that can provide employment opportunities as well as field trip experiences to allow a better understanding of marine and maritime activities. Individual museums should be contacted for brochures describing holdings and procedures. You also may wish to contact the following association:

American Association of Museums
    1055 Thomas Jefferson Street NW
    Washington, DC 20007

PILOTS

Lexie Palmore's college education in commercial art did not prepare her for her present career as a *river pilot*. Lexie serves as the pilot aboard the famous Mississippi paddleboat, the *Delta Queen*. A graduate of a training program for pilots, Lexie enjoys life on the river but feels the rigors of her demanding schedule: "You have to feel comfortable traveling all the time. I don't have a family of my own, so that isn't a problem for me."

Pilots take ships safely through ports, harbors, and rivers. By law and by custom, ships must take aboard a harbor pilot as they prepare to enter a port to guide the ship into its docking. At prearranged locations, harbor pilots meet and go aboard ships. Pilots also guide vessels out of port. For example, ships coming into the port of Philadelphia will be met by a pilot boat out of Lewes, Delaware, just as the ship passes the breakwater and enters Delaware Bay. The pilot possesses intimate knowledge of the harbor and the channel that the ships follow and keeps abreast of changes to maintain the safety of the ship and the harbor (particularly with regard to supertankers) and to avoid running aground. While pilots do give orders, the captain remains responsible for the ship and, consequently, stays in command.

Pilots serve the inland waterways as well as the major coastal harbors. The duties of a pilot on inland waters involve reporting to the captain and responsibility for two six-hour navigation watches, the safety of crew and tow while on watch, communications (navigational reports to the captain), prediction of vessel fuel utilization, and log entries while on watch.

Many years of experience aboard ships or towboats are necessary in order to become a river pilot. As with other maritime personnel, pilots must have a license. License specifics vary with the water route over which the pilot will function.

An organization that trains pilots, as needed, is the Seafarers Harry Lundenberg School of Seamanship. This school requires all applicants to show evidence of three years of service in the deck department on steam or motor vessels; eighteen months of this time shall have been as able seaman or equivalent. Of these eighteen months, one year shall have been in the capacity of quartermaster, wheelman, or an equivalent position that included standard regular watches on the wheel and in the pilothouse as part of routine duties. This service shall have been on vessels operating on the waters of the class for which pilotage is sought, and 25 percent of this service shall have been obtained during the three years preceding the application for licensing.

All candidates must have made the minimum number of normal trips over the route for which the applicants seek licensing as pilot. This minimum is determined by the OCMI (Officer in Charge, Marine Inspection) of the jurisdiction. All candidates must have made one normal trip over the route within six months immediately preceding the date of application. John Wiegman, Assistant Director at the Lundeberg School notes, "During some periods there exists a surplus of pilots, hence school training ceases."

Admission to membership in a harbor pilot's organization often is difficult. Entrance to the United New York and New Jersey Sandy Hook Pilots Benevolent Association is normally through a relative; each pilot has the right to nominate one candidate at a time. Captain Cornelius Keating, who entered the system without going through a relative, believes it is important that the job be in your blood. "It can get very dangerous.... but if you're looking forward to being a pilot [since childhood] you'll go." Harbor pilots undergo a lengthy apprenticeship, but salaries are excellent.

Some special opportunities do exist in this field. On the Delaware River, a fleet of ships (*Good Ship Lollipop, Rainbow Yacht,* and *Showboat*) take people on a tour of the harbor. These smaller versions of the old Mississippi steamboats have young people as their pilots. As high school students, they worked summers as deckhands before becoming pilots. Susan Ardary, owner of the Rainbow Fleet, explains, "They work their way up, get experience on the ship, then apply for their special license for passenger vessels from the Coast Guard. Of course, they must pass a rigid test before they earn their license."

A career as a harbor pilot permits the individual to combine an interest in ships with a permanent residence on shore.

Seafarers Harry Lundenberg School of Seamanship
    P.O. Box 75, Route 249
    Piney Point, MD 20674

WOMEN AFLOAT

In today's marine and maritime world, women participate in the full range of careers. They serve in numerous oceanography and marine science positions both in saltwater and freshwater environments. For many years, their contribution to marine ecology has been legend. For example, at North Carolina State University, Lundie Spence works with school teachers and students in her capacity as a marine science adviser. Both the educators and their students are growing in their appreciation of the marine environment.

In the maritime world, several women work aboard ships in various capacities. One of the first women to graduate from a maritime academy, Debbie Doan, works for Lykes Brothers Steamship Company. Another, Susan Janis, serves as an officer aboard the *Exxon Boston,* an oil tanker.

In Florida, Nancy Gilligan, armed with her boat captain's license, takes out fishing parties and charter groups. Additionally she transports sailboats and yachts for their owners.

On shore, Laurie Seyller serves as a sheet metal worker at Lockheed Shipbuilding in Seattle. She is just one of many women in the ship construction business.

Joy Wolf has served in numerous executive positions in the field of marine science education. She is Director of Education for Sea World in San Diego and has served as President for the National Marine Educators Association. The latter is a professional association of educators committed, "to make known the world of water, both fresh and salt."

Numerous women serve as enlisted personnel and in the officer corps for the Navy and Coast Guard. They enter through enlistment, the academies, the Officer Candidate Schools, or NROTC and work in a variety of capacities on shore and aboard ship. In the very demanding but romantic world of the sea, one's sex plays no factor in obtaining employment.

ADMIRALTY LAW

Although the world of lawyers suffers from overpopulation, one specialty with excellent promise is that of admiralty law. Following undergraduate school, lawyers must attend law school (three years) and pass the state bar examination prior to entrance into the profession. Admission to law school demands very good undergraduate grades and a good score on the LSAT (Law School Aptitude Test). Those wishing to emphasize admiralty law pursue general legal studies in law school, and in the second and third year they take courses in their special area of interest. Only a few schools offer an admiralty or a marine law concentration. One unique graduate law program, Ocean and Coastal Law, exists at the University of Miami School of Law. This curriculum requires an additional year of study in which students take courses such as the following:

admiralty
admiralty seminar
coastal law
international law
marine insurance
marine pollution seminar
ocean law
water law

Today's legal-minded society does not escape the marine and maritime community; this career offers interesting, valuable, and profitable employment.

BAIT AND FISHING SHOP OWNERS AND EMPLOYEES

Those with expertise in recreational fishing may wish to consider employment in or ownership of a bait and fishing equipment store. Obvious prerequisites are knowledge of fishing techniques, fish in season, and familiarity with new trends and equipment. Persons in these careers usually have had a lifelong interest in fishing. If you wish to own or operate a fishing and bait store, business skills and leadership ability will be necessary.

For surfboard, SCUBA, and maritime arts and crafts shops, a similar situation exists. That is, love for and experience with your product must be meshed with sound business knowledge.

BAIT-FISH CATCHERS

Millions of men and women engage in sport fishing annually. Though some of these people obtain their own bait, most do not. Consequently, catching these small fish can prove to be a lucrative and enjoyable pastime. Because of the seasonal nature of the work in many locations, another job will be necessary for the winter months. The following excerpt provides an insight into this career:

My name is Dirty Louis...I trap killies to sell to local marinas (Manahawkin, New Jersey, area). The business can be lucrative, with killies or minnows selling wholesale at good prices per quart and forty or more quarts as a decent morning's load. The only requirements for the job are a fast, flat-bottom boat, fifty to seventy-five traps, and a hell of a memory of all the mainland creeks, bay islands, and Sheeps head Peninsula waterways. The secret to trapping killies is to set up in low water and catch the fish as they come into the creeks, but before they can get into the flooded grass with high tide and get other food. Conical traps are baited with bunker, an oily fish, and, with the lack of other eatables in the low creek, the minnows enter the trapping pot. The traps are left in the creeks for one-half to one and one-half hours, depending on location, then pulled and the trapped minnows are transferred to floating boxes that keep them alive...

Trapping generally starts on the warmer mainland streams in June and early July, then switches to the bay islands in July and then Sheepshead Peninsula near the inlet in August. The same stream should be trapped only once every week or two; thus, a broad knowledge of streams is required. Each trap pulled is a test of one's ability to judge the tide, determine the correct creek, and estimate where in the creek traps are best placed. As the trap quietly breaks the surface, the sounds of tiny slapping minnows crammed in the wire mesh signal a worthwhile catch.

Bait fishing varies widely with geography, temperature, nature of the fish for which the bait will be used, and other factors. On the Atlantic coast, the commercial bloodworm and sandworm can constitute a valuable enterprise. One thing remains constant—this old craft can provide a happy, healthy, and financially worthwhile vocation or avocation.

MARINE INSURANCE CAREERS

All ocean-related industries require insurance; ships, off-shore platforms, cargoes, yachts, marinas, shipbuilding companies, and similar businesses insure themselves against loss. This is not a recent phenomenon. The Insurance Company of North America (now CIGNA), founded in 1792, issued its first policy to cover the *American,* sailing from Philadelphia to Londonderry (Ireland).

A typical person entering the marine insurance field at CIGNA (usually a college graduate) would spend the first six months learning general insurance and take correspondence courses on marine insurance. After that, the employee would attend the company's formal Marine and Aviation School, approximately four weeks long. Following school, individuals receive their assignments and begin their careers selling insurance, investigating claims, providing safety programs, planning cargo loss control programs, arranging claims settlements, conducting safety inspections, and performing similar activities.

For additional information, contact the nearest office of CIGNA.

SEAFOOD PROCESSING AND MARKETING CAREERS

A surprising number of people work in the seafood processing and marketing industry. Presently, more than 150,000 persons process the catches of fishermen, and this number increases yearly. The industry is unique not only because of the perishable nature of the fish but because of regional eating preferences and sources of fish.

For example in North Carolina the methods for marketing seafood have changed little over the last several years. The fishermen sell their catch to local packers where the fish, upon icing and packing, are sold to retail markets as far away as New York. Some processors freeze the fish and sell it as a frozen food. Traditionally, North Carolina has focused on the sale of whole fresh finfish and shellfish.

Close study of fishing in the tar-heel state seems to indicate that new ideas will modify the industry. The development of seafood-related industrial parks at key locations along the coast, if properly implemented, will increase the state's processing capacity. Present plans call for expansion of the market into the Midwest. The region, chosen for its potential and proximity to North Carolina, includes eighteen target cities in Minnesota, Wisconsin, Iowa, Missouri, Illinois, Indiana, Michigan, Ohio, Tennessee, and Kentucky. The promotion will feature croaker, gray trout, king and Spanish mackerel, mullet, and rock shrimp. Clearly this campaign will require successful selling techniques. In each city the promoters visit buyers, offer samples, and provide information. Marketing specialists meet with newspaper food editors as well as television and radio program directors, distribute brochures, and generally try to educate the public about the qualities of the new species. Such an undertaking, though costly at first, will often pay off in later success.

The dramatic increase in aquaculture (fish farming) has greatly affected processing and distribution. For example, New Jersey has begun an oyster farm project in the Cape May area. Billed as "Camp May Salt Oysters," these disease-resistant oysters are marketed not only in the greater New Jersey–Philadelphia area but also to New England.

Consumption of fish in the United States expands annually and predictions forecast continued growth. New seafood restaurants open, seafood menus grow, and home use is up. Norm Angel, executive secretary of the North Carolina Fisheries Association, explains his feelings: "If I were to go into fish marketing today, I would beg, borrow, and steal all the money I could to buy myself a wholly owned freezer and custom cut the fish for the consumer market.... this is where the money is."

ACADEMIC CAREERS

A number of careers exist in colleges, universities, and technical institutes as *teaching* and *research faculty.*

Individuals in these occupations teach courses ranging from oceanography and maritime history to literature of the sea and trends in aquaculture. Those in these university careers have strong educational backgrounds normally possessing master's and doctoral degrees. Those teaching technical courses have significant experience in their field.

These individuals enjoy the marriage of teaching while still participating in the practical world of water. In addition, they possess intellectual curiosity pursuing new knowledge through research and development.

Those interested in careers related to maritime studies should contact:

The American Neptune
  Publications Department
  Peabody Essex Museum
  East India Square
  Salem, MA 01970-3783

Munson Institute
  Mystic Seaport Museum
  P.O. Box 6000
  Mystic, CT 06355-0990

Naval Historical Foundation and
  Naval Historical Center
  Build 57 Washington Navy
  Yard
  901 M Street SE
  Washington, DC 20374-5060
  www.history.navy.mil

North American Society for Oceanic History
Department of History
East Carolina University
Greenville, NC 27858-4353

Additionally, the following web site entitled Maritime Education provides a sample of the range of careers in this field: http://ils.unc.edu/maritime/mared.html.

A number of other unique but limited careers exist in the marine and maritime world. These run from a *smokehouse operator* (where fish are smoked) to *underwater photographer.* Those with an interest should read related publications and maintain contacts to remain abreast of career developments in these exciting fields.

CHAPTER 10

GETTING YOUR FEET WET

Many of the careers listed in the book offer personally and professionally rewarding opportunities. However, at this point many people need more information and some hands-on experience.

Reading can open a world of knowledge (see Suggested Readings in Appendix A). Organizations in your field of interest offer excellent information. In addition, involvement activities such as an internship are a must. Some suggestions follow.

PRE-CAREER EXPERIENCES

Visits

Aquaria, nature centers, marine sanctuaries, and maritime museums offer wonderful awareness and in some cases hands-on opportunities. Many of these institutions exist in our country, so that a visit is not too far for anyone. Maritime museums offer excellent possibilities for learning experiences. Likewise, many fine aquaria exist (see Chapter 1). For example, the New England Aquarium, located in Boston, provides excellent programs, including multimedia shows, observation of divers feeding fish, touch-and-see aquatic animals, dolphin and sea lion shows, a Sea Explorer post, and independent study. Most importantly, many aquaria have specialized programs and volunteer opportunities for young people. For example, the Vancouver Aquarium boasts a Youth Volunteer Team designed "to build future leaders in conservation and career skills;" consult the Youth Volunteer Lounge link on the aquarium's web page: www.vanaqua.org/volunteer/volunteer.htm. Organizations such as Sea World offer similar activities specifically for young people. The San Antonio facility offers a Careers Camp for high school students. A week in length, "this once-in-a-lifetime opportunity permits daily interaction with all aspects of marine life." They also offer an Advanced Career Camp. To participate in such programs, contact the Aquarium/Oceanarium's Education Department. For additional opportunities, you may wish to check the web site of the National Association of Marine Laboratories (NAML) at www.mbl.edu/html/NAML/NAML.

School Programs

If you are a student, ask your teacher to plan marine and maritime studies field trips and encourage your school to integrate these topics into the curriculum. Look into all available marine and maritime courses and programs. Many examples of the types and content of school courses are detailed in this book. Volunteer to aid in setting up the program. Approach faculty members and ask them to sponsor a marine science, SCUBA diving, or maritime careers club; or originate an NJROTC program. These experiences can be profitable for educators and students. You may take field trips, invite speakers, have discussions, and organize exhibits at the school and in the community. You may even get to take a virtual field trip on the web; the Jason Project hosts an excellent site at www.jasonproject.org.

Opportunities

Several businesses and schools regularly schedule cruises combining recreation and education for young people. Some of these are the Sea Education Association (SEA); Schooner, Inc.; the University of Hawaii's Blue Water Marine Laboratory; the Summer School at Sea of Texas A & M; the Institute for Shipboard Education (Semester at Sea); and Marine Sciences Under Sails.

These typical programs offer college credit and have objectives similar to SEA's Sea Semester, which is a rigorous academic, scientific, and practical experience leading to a thorough understanding of the sea. The Shipboard Education voyage resembles a luxury cruise in which students take courses and visit United States and foreign ports while enjoying life aboard ship. The college cruises vary in sophistication. Advanced high school students may participate in some programs.

Schooner, Inc., operating on the Long Island Sound, has conducted many fine experiences for teenagers from the greater New York City area. In a program focused on the environment, one student, John Leeds, from Choate/Rosemary Hall, discovered, "My God, the slime is really horrible." A classmate observed, "It's like people telling you all about the starving people in India; you really don't know until you see it."

The precollegiate cruises generally have an objective similar to one of Hawaii's programs: The primary purpose of the Blue Water Marine Laboratory Program is to increase the awareness, knowledge, and understanding of Hawaii's marine environment in the intermediate and secondary school population in the state, and to do so by offering unique seagoing experiences in ocean studies. Most of the learning activities involve students spending approximately four hours obtaining samples of water and aquatic species, experiencing life aboard ship, and investigating life along the shoreline.

For interesting experiences have your school contact the following:

American Tall Ship Education
    Association
    One Bostwick Avenue
    Bridgeport, CT 06605

Blue Water Marine Laboratory
    George Hall 230
    2560 Campus Road
    Honolulu, HI 96822

Dirigo Cruises
    39 Waterside Lane
    Clinton, CT 06413

Institute for Shipboard Education
    Taj Mahal Building, Suite 203
    P.O. Box 2488
    Laguna Hills, CA 92653

Marine Sciences Under Sails
    P.O. Box 2994
    Hollywood, FL 33023

Ocean Challenge
    20 Park Plaza
    Suite 424
    Boston, MA 02116

Ocean Schools Foundation
    25 Woodman Road
    Durham, NH 03824

The Propeller Club of the United
    States
    3927 Old Lee Highway, Suite
    10A
    Fairfax, VA 22030

Sea Education Association
    P.O. Box 6
    Woods Hole, MA 02543

Schooner, Inc.
    60 South Water Street
    New Haven, CT 06519

Summer School at Sea
    Texas A & M University
    P.O. Box 1675
    Galveston, TX 77553

If your college does not offer marine science courses, consider a summer program that does. The University of Hawaii (at Manoa) offers World Maritime History and Maritime Archaeology Techniques. Southampton College of Long Island University plans an excellent selection of courses: Underwater Photography, Marine Ecology, Traditional Maritime Culture, and others.

Think about becoming a student member of the National Marine Education Association; its publications frequently contain lists of summer opportunities in the marine sciences.

National Marine Education
    Association
P.O. Box 1470
Ocean Springs, MS 39566-1470

To remain current in the maritime world, the following organizations will help:

National Maritime Historical
    Society
P.O. Box 68
Peekskill, NY 10566
www.seahistory.org

North American Society for
    Oceanic History
c/o Department of History
East Carolina University
Greenville, NC 27858-4353

In addition to scanning the newspapers for commercial advertisements, you can find out more about learning experiences aboard ships by consulting Sea Grant educational advisors at colleges and universities offering this program (see Chapter 3).

A number of organizations offer boating/sailing training programs: The Chapman School of Seamanship in Stuart, Florida, schedules Summer Boating for Youth (ages eleven to seventeen); Boat Camp (ages thirteen to eighteen) in San Diego forms part of the Discover the Ocean, Inc., program; and Kid Ship (ages five to fifteen) is located at the Annapolis Sailing School.

Often a river or harbor cruise can increase your awareness of career activities; almost every city has one. Contact the following:

Grayline Sight-Seeing Association
    7 West Fifty-first Street
New York, NY 10019

Those interested in maritime transportation will want to look at the following web page, Especially for Kids and Teens: www.marad.dot.gov/KIDS.html. It contains a number of fascinating links.

SEA CADETS

The Navy League of the United States, a private organization of individuals interested in the Navy, sponsors two youth programs. Their activities are aimed at educating and training young people in citizenship and maritime matters, exposing them to naval history and tradition, educating them in the importance of maritime forces to our nation, and explaining to them the opportunities available in maritime careers.

There are two coeducational programs: the Naval Sea Cadet Corps (NSCC) for ages fourteen to seventeen and the Navy League Cadet Corps (NLCC) for ages eleven to thirteen. There are approximately two hundred NSCC and NLCC units in the United States and Puerto Rico with more than seventy-five hundred officers and cadets involved. Sea Cadets receive their training aboard ships, planes, and in shore installations of the Navy and Coast Guard. In addition, they are authorized to attend a two-week recruit training experience. Graduates are eligible for advanced training through special shipboard programs held during school holidays and summer months.

Programs for younger people (NLCC) do not include training aboard Navy or Coast Guard ships or aircraft. The education covers naval history, customs and traditions, water safety, first aid, leadership procedures, and physical training. Several field trips to Navy and Coast Guard facilities complement the lectures. Both Sea Cadets and Navy League Cadets wear modified Navy uniforms.

Membership in the Sea Cadets permits firsthand experience with the Navy and Coast Guard. Cadets who decide to enlist in the Navy may be eligible to do so at an advanced pay grade based on their cadet records. A scholarship fund is available for those interested in higher education. If this program appeals to you, contact the Sea Cadets group nearest to you or write the following:

Naval Sea Cadet Corps
2300 Wilson Boulevard
Arlington, VA 22201

SEA EXPLORERS

The exploring program of the Boy Scouts of America works for the improvement of the individual in these areas: vocation, personal fitness, citizenship, social service, and the outdoors. Specifically, the Sea Explorers operate activities such as the following:

The National Sea Explorer Sailing Championships is a program that originates with local councils, involves area and regional sail-offs, and climaxes with national competition under the supervision of the United States Racing Union.

Sea Explorer Rendezvous, usually held on a council, area, or regional basis, gives the Explorers a chance to demonstrate through competition many of the traditional, yet practical, skills of seamanship.

Oceanography Seminars are growing in popularity. These are usually conducted in cooperation with federal and state agencies, universities, and similar research organizations.

Cruising is popular with Sea Explorer ships. Some cruises are held by individual ships—others involve several ships operating as a flotilla. Rivers, lakes, coastal waterways, and the open ocean are typical cruising sites. Often long-term or short-term cruises are available through the United States Navy or Coast Guard.

High Adventure Gateways, operated by the Boy Scouts of America, are available to Sea Explorers. Of particular interest is the Florida High Adventure Gateway offering underwater exploration, skin and SCUBA diving, and sailing in the Florida Keys. The Land Between the Lakes Gateway in western Kentucky and Tennessee also offers experiences in sailing and power boating.

Other areas of interest are found in *Traditional Sea Exploring,* which centers on the sea, life and duties aboard ship, and maritime customs. In some areas this program is known as the *Blue Fleet. Careers in Inner Space* forms the basis for activities in the marine sciences, fisheries, oceanography, marine architecture, and others. Persons in this program, often called the *Red Fleet,* frequently work with professionals in the careers they wish to enter. The exciting leisure activities of aquatics, sailing, boating, and SCUBA, form a part of sea exploring; in some regions this program is entitled the *White Fleet.*

Young adults, ages fifteen to twenty-one, may join Sea Exploring. Most ships are coeducational; however, some are all male, others all female. There are more than fifteen hundred Sea Explorer units with over twenty-one thousand young people involved. Many groups own their own vessels. If you wish additional information, write the following:

Sea Exploring
  1325 Walnut Hill Lane
  Irving, TX 75038

NAVAL JUNIOR RESERVE OFFICER
TRAINING CORPS (NJROTC)

This Navy program offers the opportunity for secondary school students to learn about the Navy, national security, and marine and maritime topics. The curriculum runs either three or four years, depending upon the school. One course is taken each year. Naval Science III, taken during the third or fourth year, emphasizes the following concepts:

America's maritime heritage
  drill, commands, and ceremonies
  health education
  leadership
  navigation

oceanography
orientation of sea power
seamanship
shipboard indoctrination

In addition to classroom instruction, participation in the drill team (which is part of all units), the band (available at many units), and the rifle team (available at some schools) is encouraged. Those involved in these activities engage in competitive matches with other units. Field trips and cruises form an exciting part of the program of all units. Inland units are often airlifted to coastal installations.

More than two hundred NJROTC units exist with approximately thirty thousand male and female high school students participating. Students receive one unit of credit for each year of course work. Uniforms are similar to those of active duty personnel, and special badges and insignia are provided by the Navy. If your school has this program, contact one of the instructors for specifics. If not, you may be permitted under certain conditions to affiliate with a high school that does. Also, you may wish to work to establish a unit. Write to the Navy, addressing your letter to:

NJROTC
    CNET
    250 Dallas Street
    Code OTE10
    Pensacola, FL 32508-8520

THE GIRL SCOUTS

Some of the finest world-of-water programs available for young people are integrated into the Girl Scouts programs for ages twelve to seventeen. Nancy Richardson, Program Specialist in Marine Education, reports the activities resulted as a response, "to those senior Girl Scouts who love the water and are eager to know more of the fascinating life of the sea and all its ramifications."

The chapters, called Ships, vary widely. The Marina Scout Ship (MSS) *Shangri-La* in Los Altos, California, has built more than one hundred kayaks in the last several years. Based on its own success, the group has offered advice on similar programs to the Boy Scouts, 4-H, and Campfire Girls groups. (Some of the chapters of the latter two organizations also offer maritime and marine activities for their members; contact your local group.) Mariners also engage in ship restoration. The *Alma,* located in San Francisco, the *Adventures* in Seattle, and the *Joseph Conrad* in Mystic have benefited from their care.

Several councils offer summer opportunities in cruising. The San Jacinto group of Houston describes their trip: "Long to see the world? Then 'come aboard' and experience the Texas Gulf coast as no landlubber ever has." Not to be outdone, the Toten Girl Scout Council of Seattle offers the opportunity to, "Hear the wind, the sails, share your dreams with a seagull, stand watch, cook in a gimballed galley, sleep at anchor, build a sand castle, and sing chanties." The Napa-Solano chapter of Vallejo, California, has organized The Trip Around the Bay. Participants "awake each morning at the California Maritime Academy to a glistening view of the Carquinez Straits—fastest flowing waters of the bay." Then they travel to different learning sites to experience the plant and animal life of the marshes and water.

Other programs—Canadian Canoe Trails, Finger Lakes Sailing Festival, Hellhole Swamp Stomp, and Spree by the Sea—offer wonderfully enjoyable and educational adventures. Girl Scouts from all over the country travel to participate in them. Most activities are open to young women only. However, some are coed and others are held in cooperation with Sea Explorers. In addition to the cruises available nationally, numerous ships and local councils for ages six to twelve engage in marine and maritime studies. For more information contact a local group (local councils are listed in the telephone book).

Career immersion in the marine and maritime specialties often involves more than wet feet!

SUGGESTED READINGS

PERIODICALS

American Neptune. A research publication of the Salem Peabody Museum. Containing interesting articles of a scholarly nature, this journal provides excellent information on America's maritime history. Peabody Museum, East India Square, Salem, MA 01970.

Boating. A popular, nicely illustrated magazine containing historical, contemporary, and how-to articles. Students will find advertisements as well as articles of interest. Located at 1633 Broadway, Forty-third Floor, New York, NY 10009.

Cruising World. A popular magazine aimed at readers interested in sailing. Each issue contains articles on technique, possible adventures, tours, and other general sailing–related topics. The illustrations and advertisements will be enjoyed by students. Sailing Company, Box 3400, Newport, RI 02840.

Current: The Journal of Marine Educators. The well-written periodical of the National Marine Educators Association. Contains popular articles as well as updates of marine science and maritime affairs. National Marine Education Association, P.O. Box 1470, Ocean Springs, MS 39566-1470.

Dolphin Log. A marine science–oriented publication for younger children. This is a very well-written magazine from the Cousteau

Society. Located at 870 Greenbrier Circle, Suite 402, Chesapeake, VA 23320.

The Log of Mystic Seaport. A well-done research-oriented publication. The focus is upon maritime history. Mystic Seaport, Mystic, CT 06355.

Marine Conservation News. A newsletter filled with updates of all aspects of marine life. Center for Marine Conservation, 1725 DeSales Street NW, Washington, DC 20036.

Marine Mammal Science. For those with a strong interest in this field. Dr. Glen Van Blaricom School of Fisheries, University of Washington, Seattle, WA 98195.

Nor'Easter. A well-done publication of the Northeast Sea Grant programs. It contains updates on research related to the sea as well as updates on NE regional programs. Each of the Sea Grant colleges regularly publishes such a journal. Another example is the *University of Delaware's Sea Grant Reporter.* Contact your regional college (see Chapter 3) for a subscription.

Oceanus. A quarterly publication investigating and reporting recent findings on the sea. Topics will interest persons with a concern for the frontiers of sea research. Woods Hole Oceanographic Institute, 360 Woods Hole Road, Woods Hole, MA 02543.

Professional Mariner. Published six times a year, this magazine focuses upon navigation and seamanship. Navigator Publishing, 18 Danforth Street, Portland, ME 04101.

SCUBA Diving. A monthly magazine lavishly illustrated focusing upon all aspects of SCUBA from equipment to interesting dives. Rodale Press, 33 East Minor Street, Emmaus, PA 18098.

Sea Frontiers. A magazine with illustrated articles concerning advances, explorations, and discoveries related to the marine sciences. Published by International Oceanographic Foundation, 400 SE Second Avenue, 4th Floor, Knight Centre, Miami, FL 33131. This publication incorporates *Sea Secrets.*

Sea Heritage News. A newspaper with a wide variety of maritime-humanities articles (art, language, history). Sea Heritage Foundation, 254-26 Seventy-fifth Street, Glen Oaks, NY 11004.

Sea History. Clearly one of the finest periodicals in the field. Each issue contains historical articles as well as current issues of historical interest. Additionally, maritime art receives attention. National Maritime Historical Society, 5 John Walsh Boulevard, Box 68, Peekskill, NY 10566.

Sea Power. A publication of the Navy League, an independent, nonprofit, civilian education organization. Recent editions have contained illustrated articles on various aspects of maritime life—fishing industry, oceanography, and shipbuilding. Navy League, 2300 Wilson Boulevard, Arlington, VA 22201.

Steamboat Bill. A quarterly publication of the Steamship Historical Society of America. This journal contains short, interesting, well-illustrated features on topics such as ferryboats, Mississippi showboats, and other nonnaval steamships. Located at 300 Ray Drive, Suite 4, Providence, RI 02906.

United States Naval Institute Proceedings. A publication of the United States Naval Institute, a private, professional society for all who are interested in naval and maritime affairs. Past issues of this more than 125-year-old journal constitute an excellent source of maritime (especially naval) history. More recent issues deal with contemporary maritime and naval problems. The institute also publishes books related to naval and maritime affairs. *Naval History,* published bimonthly, provides quality articles on a wide range of popular and less-known aspects of naval history. United States Naval Institute, 118 Maryland Avenue, Annapolis, MD 21402.

Wooden Boat. A bimonthly magazine on present-day building, repair, and restoration of wooden craft. *Wooden Boat,* Box 78, Brooklin, ME 04616.

MARITIME MUSEUMS

Alabama

Dauphin Island: Fort Gaines Museum, USS *Hartford*
Gulf Shores: Fort Morgan Museum, lighthouses
Mobile: USS *Alabama*, USS *Drum,* USS *Tecumseh*

Alaska

Juneau: Relics of USS *Juneau*

California

Dana Point: Nautical Heritage Museum
Long Beach: SS *Queen Mary*
Monterey: Allen Knight Maritime Museum, The *San Diego*
Port Hueneme: Seabee Museum
San Diego: Maritime Museum of San Diego
San Francisco: San Francisco Maritime National Historical
    Park, Schooner C.A. *Thayer,* Steam-Schooner *Gjoa,* others;
    Treasure Island Museum
San Pedro: Los Angeles Maritime Museum

Connecticut

Bridgeport: HMS *Rose*
Essex: The Connecticut River Museum, The *American Turtle*
Groton: *Nautilus* Memorial

Mystic: Marine Historical Association and Mystic Seaport, whaleship *Charles W. Morgan*, ship *Joseph Conrad*, schooner L.S. *Dunton*, sandbagger *Annie*, cutter *Fox*, sloop *Estella A.*, schooner *Bowdoin, Dorothy Parsons, Regina M.*, schooner *Pleione, Gundel*

New London: United States Coast Guard Museum, The *Eagle* and others

Delaware

Christiana: A number of ships under restoration

Lewes: Cannon Ball Marine Museum; University of Delaware Marine Science Center

District of Columbia

Santa Maria; Smithsonian Institution, Hall of American Maritime Enterprise; Navy Museum; United States Naval Historical Display Center, Washington Navy Yard; Washington National Archives

Florida

Fort Lauderdale: *Jungle Cruise,* Paddleboat

Key West: Mel Fisher Maritime Heritage Society

Marathon: Sunken Treasure Ship Museum

Pensacola: Naval Aviation Museum

St. Petersburg: HMS *Bounty* (restoration)

Georgia

Atlanta: HMS *Bounty* exhibit

Columbus: Confederate Naval Museum; CSS *Muscogee,* CSS *Chattahoochee*

Savannah: Model of SS *Savannah* at City Hall; Ships of the Sea Maritime Museum, *Flying Cloud* (eight-foot model), *Harbor Queen*

Hawaii

Honolulu: USS *Arizona* Memorial, Pacific Submarine Museum

Illinois

Chicago: Chicago Maritime Society; Museum of Science and
    Industry, Submarine U-505 (German, World War II); replica
    of Viking ship in Lincoln Park
Lockport: Illinois and Michigan Canal Museum
Peoria: *Belle Reynolds*

Indiana

Jeffersonville: Howard Steamboat Museum

Iowa

Dubuque: Fred W. Woodward Riverboat Museum
Keokuk: Keokuk River Museum, *George M. Verity*

Kentucky

Louisville: *Belle of Louisville*

Louisiana

Baton Rouge: Louisiana Naval War Memorial
New Orleans: CSS *Pioneer;* Louisiana State Museum

Maine

Bath: Maine Maritime Museum; *Sherman Zwicker;* wooden
    boatbuilding
Castime: Allie Ryan Maritime Museum
Portland: Lightship *Nantucket*
Searsport: Penobscot Marine Museum

Maryland

Annapolis: Historic Annapolis, Inc., Slicer Shiplap House;
    United States Naval Academy Museum; United States
    Naval Academy Library

Baltimore: Maryland Historical Society, Marine Wing; USS
Constellation; the Steamship Historical Society of America
Library; Inner Harbor Baltimore Maritime Museum; The
Chesapeake, USS *Torsk;* Baltimore Museum of Industry,
The *Baltimore;* Calvert County Marine Museum
Cambridge: Bannock Maritime Museum
Chesapeake City: C and D Canal Museum
Landover Hills: Nautical Research Guild
Piney Point: *Dauntless,* Harry Lundeberg Schools of Seaman
ship, *Manitou*
St. Michaels: Chesapeake Bay Maritime Museum, Hooper
Strait Lighthouse

Massachusetts

Andover: Addison Gallery of American Art, Phillips Academy
Barnstable: The Donald G. Trayser Memorial Museum
Boston: Boston Marine Society; Museum of Fine Arts; Old
State House; Museum of Science, Science Park; State Street
Trust Company; Stebbins Marine Collection, Harrison Grey
Otis House; USS *Constitution,* Boston Tea Party Ship and
Museum
Cambridge: Francis Russell Hart Nautical Museum, Massachu-
setts Institute of Technology
Chatham: Whaling Museum
Cohasset: Maritime Museum
Edgartown: Dukes County Historical Society, Martha's Vine-
yard
Fall River: Fall River Historical Society Museum; USS *Massa-
chusetts,* State Pier, USS *Lionship;* USS *John Kennedy,* USS
Lionfish
Gloucester: Cape Ann Historical Museum, Gloucester Fisher-
man's Museum
Milton: Museum of the American China Trade

Nantucket: Nantucket Whaling Museum, Nantucket Lifesaving
   Museum
New Bedford: Whaling Museum and Old Dartmouth Historical
   Society
Plymouth: *Mayflower II* (replica)
Salem: Peabody Museum, 161 Essex Street; Salem Maritime
   National Historic Sites; Custom House; Essex Institute
Sharon: Kendall Whaling Museum

Michigan

Detroit: Detroit Historical Society; Dossin Museum of Great
   Lakes History, *Belle Isle*
Mackinac Island: Mackinac Maritime Park
Menominee: Mystery Ship Seaport
Sault Ste. Marie: Museum ship *Valley Camp,* Great Lakes
   freighter
South Haven: Lake Michigan Maritime Museum

Minnesota

Winona: Transportation Museum on board the *Julius C. Wilkie*

Mississippi

Biloxi: Seafood Industry Museum
Vicksburg: USS *Cairo* Museum

Missouri

Hannibal: Steamer Mark Twain
Hermann: Historic Hermann Museum, including Riverboat
   Room
St. Louis: Missouri Historical Society, River Room, *Golden
   Rod,* USS *Inaugural*

Nebraska

Brownville: Museum of Missouri River History

New Hampshire

Portsmouth: Port of Portsmouth Maritime Museum, library, paintings, ship models

New Jersey

Barnegat Light: Historical Society Museum, typical of many of the museums located along the New Jersey coast; Barnegat Lighthouse

Beach Haven: home of the Long Beach Island Historical Society

Hackensack: US Naval Museum, USS *Ling*

Hoboken: Hudson Waterfront Museum, the *Philip T. Feeney*

Paterson: City of Paterson Museum, *Fenian Ram*

Sandyhook: Sandyhook lighthouse

New York

Amagansett: Town Marine Museum

Amsterdam: Fort Hunter, Erie Canal Locks

Brooklyn: Brooklyn Museum

Buffalo: Buffalo and Erie County Naval and Serviceman's Park

Clayton: Thousand Islands Shipyard Museum

Cold Spring Harbor, Long Island: Vanderbilt Marine Museum

East Hampton: East Hampton Town Marine Museum

Glen Cove, Long Island: Webb Institute of Naval Architecture

High Falls: Delaware and Hudson Canal Museum

Huntington: Whaling Museum Society

Hyde Park: Frankling D. Roosevelt Library and Museum

Kings Point: American Merchant Marine Museum, National Maritime Hall of Fame

Kingston: Hudson River Maritime Center, The *Mathilda*

Lake Champlain: Fort Ticonderoga Museum

New York City: India House, Hanover Square; Metropolitan Museum of Art; Museum of Science and Industry, Seamen's Church Institute, New York Public Library, Fifth Avenue at

Forty-fourth Street; South Street Seaport, *Ambrose* light-
ship, *Lettie G. Howard, Howard, Alexander Hamilton;*
Intrepid, Sea-Air-Space Museum, USS *Intrepid,* USS
Growler, USS *Edson*
Sag Harbor: Suffolk County Whaling Museum
Staten Island: Staten Island Museum, The Steamship Historical
Society of America Library; National Lighthouse Center
and Museum
Syracuse: Erie Canal Museum
West Sayville: Suffolk Marine Museum

North Carolina

Beaufort: North Carolina Maritime Museum (This museum
conducts a number of programs of interest to the public and
for those with an interest in marine or maritime careers.)
Kinston: CSS *Neuse,* Governor Richard Caswell Memorial
Manteo: Elizabeth II State Historic Site
Wilmington: USS *North Carolina* Memorial

Ohio

Canal Fulton: Canal Fulton Heritage Society Museum,
St Helena II
Cleveland: Great Lakes Historical Society
Fairport Harbor: Fairport Marine Museum
Marietta: Ohio River Museum, Steamboat *W. P. Snyder*
Vermillion: Great Lakes Historical Society Museum

Oklahoma

Muskogee: Muskogee War Memorial Park and Military
Museum

Oregon

Astoria: Columbia River Maritime Museum
Newport: Old Yacquina Bay Lighthouse
Portland: Oregon Historical Museum

Pennsylvania

Doylestown: Marine Room, Bucks County Historical Society
Erie: Erie Historical Museum and Planetarium
Philadelphia: Franklin Institute; Independence Seaport
Museum; Titanic Historical Society Collection, Atwater
Kent Museum; CIGNA Museum; Penn's Landing; USS
Olympia; Welcome (replica of William Penn's ship);
Gazela; Mosulu; Barnegat

Rhode Island

Bristol: Herreshoff Marine Museum
East Greenwich: Varnum Military and Naval Museum
Newport: Museum of Yachting, *Shamrock V,* Naval War Col-
lege Museum
Providence: Seaport '76 Foundation, *The Providence* (replica)

South Carolina

Charleston: Patriot's Point Maritime and Naval Museum, USS
Yorktown, NS *Savannah,* and others

South Dakota

Sioux Falls: Battleship *South Dakota*

Tennessee

Jacksonville: USS *Key West,* USS *Tawah,* USS *Elfin* (restoration planned)
Memphis: Mississippi River Museum at Mudd Island

Texas

Corpus Christi: USS *Lexington* Museum on the Bay
Fredericksburg: Fleet Admiral Chester W. Nimitz Historic Park
Galveston: The *Elissa*
Rockport: Texas Maritime Museum

Vermont

Shelburne: Shelburne Museum, Steamer *Ticonderoga* (paddle wheeler)

Virginia

Alexandria: Alexandria Seaport Foundation, Alexandria Waterfront Museum
Jamestown: *Discovery, Goodspeed,* and *Susan Constant*
Newport News: The Mariners Museum
Norfolk: Hampton Roads Naval Museum, USS *Franklin* (bridge of ship), SS *United States*
Portsmouth: Portsmouth Naval Shipyard Museum, Portsmouth Coast Guard Museum, *Portsmouth* (lightship)
Virginia Beach: Cape Henry Lighthouse, Maritime Historical Museum
Williamsburg: Jamestown Settlement
Yorktown: Colonial National Park Museum

Washington

Bremerton: Bremerton Naval Museum
Edmonds: Edmonds Historical Museum
Keyport: Naval Undersea Museum
Seattle: The Center for Wooden Boats, Coast Guard Museum Northwest

Wisconsin

Manitowoc: Manitowoc Maritime Museum, USS *Cobia*
Superior: Head of the Lakes Maritime Society

Likewise, Canada has a number of fine museums with a marine/maritime flavor, such as the Marine Museum of the Great Lakes in Kingston and the Marine Museum of the Atlantic in Nova Scotia.